# THE COMPLETE BOOK OF BRITISH BERRIES

*The breris with thaire beries bent over the wayes*
*As honysoucles hongyng uppon eche half,*
*Chesteynes and chiries that children desiren*
*Were loigged undre leves ful lusty to seen.*
*The hauthorne so holsum I beheulde eeke,*
*And hough the benes blowid and the brome-floures;*
*Peris and plummes and pesecoddes grene......*
*And other fruytz felefold in feldes and closes,*
*To nempne alle the names hit nedith not here.*

FROM 15TH CENTURY FRAGMENT LINES 898-904, 908, 909. MS BRIT.MUSADD.
E.E.T.S.(1936) ED. DAY & STEELE

'MUM AND THE SOTHSEGGER'

# THE COMPLETE BOOK OF BRITISH BERRIES

**David C Lang**

LINE DRAWINGS BY
PATRICIA DONOVAN

THRESHOLD BOOKS

First published in 1987
by Threshold Books Ltd,
661 Fulham Road,
London SW6 5PZ



Designed by Gwyn Lewis

Typeset by Falcon Graphic Art Ltd,
Wallington, Surrey
Printed and bound by arrangement with
Graphicom, Vicenza, Italy

*British Library Cataloguing in Publication Data*
Lang, David,
The Complete Book of British Berries.
1. Botany—Great Britain
2. Berries—Great Britain—Identification
I. Title
582'.0464 QK306

ISBN 0–901366–34–X

# · CONTENTS ·

# · ACKNOWLEDGEMENTS ·

In the course of study and research for this book on fruits I have been helped by experts in botany and other fields, and all have been generous with their advice and guidance on aspects where their knowledge and experience were vastly superior to my own.

I am particularly grateful to Pat Donovan for the painstaking care she has taken with the drawings in this book; they are an excellent tribute to her combined skills both as artist and botanist. My especial thanks also to those who have guided me with the more difficult genera: the Rev G.G. Graham with Roses, Dr Alan Newton with *Rubi*, Jennie Page with *Ruscus*, Alison Rutherford with Ivies, Alan Stirling with Roses and *Rubi*, and Alfred Slack for help with montane species. I am also indebted to Dr Newton for his permission to reproduce the bramble map.

In the field of human toxicology I am very grateful to Dr Glyn Volans and A.P. Whitehead, MSc., of the Poisons Unit, New Cross Hospital, for the substantial help they have provided in researching and updating the sections on the poisonous species.

Several species fruit far better in Scandinavia than in Britain, and I send my sincere thanks to Rolf Jacobson and Dr Erik Ohlsson for the slides and material which they so kindly sent me, as did the late Rolf Dalén.

My thanks are due also to Frances Abraham, Mary Briggs, Eric Clement, Kate Cosway, Dr Jennifer Edmonds, Stephen Evans, Lady Rosemary Fitzgerald, George Forster, Dr Oliver Gilbert, Chris Haes, Dr Quentin Kay, the Director and staff of the Royal Botanic Gardens, Kew, Jill Lucas, Dr Charles Nelson, Dr Michael Proctor, Dr Francis Rose, Julian Scott, Dr David Shimwell, Eric Swann, Maureen Turner, Michael Wilkinson and many other kind friends.

# · INTRODUCTION ·

Berries and fruits have always fascinated me by virtue of their shapes and colours, which brighten woods and hedgerows at a time of year when the summer flowers are fast disappearing. Each autumn they fill the country with an abundance which seems even more impressive than that remembered from previous years.

Most botany books describe the leaves and flowers of plants in precise detail, illustrating the flowers, but consigning a description of the fruits to a few brief notes. Late summer and autumn are among the finest times of the year for walking in the countryside, and this book was conceived to assist the naturalist with minimal botanical knowledge to identify plants in fruit from which the flowers and even the leaves may have gone.

The plants illustrated bear berries or fleshy fruits, such as drupes and pomes, and the list contains all such plants native to the British Isles, as well as those introduced species which have become so well established that they are part of the accepted botanical scene. In choosing the latter there is an element of selection for which I alone am responsible; the list of alien plants with such fruits is large, and most of them persist only in small areas and for a limited time.

The history of social uses of the fruits are described, and, where the fruit is toxic for man or animals, there is a description of the symptoms of poisoning and the first aid treatment which should be given.

Among the berries and fruits illustrated are some which are beautiful and fascinating, some which are useful, and others which are dangerous. This book will help you to identify those that you find, and will, I hope, widen your enjoyment of the summer and autumn countryside wherever you may be.

*David Lang*

LEWES, 1987

# · HOW TO USE THIS BOOK ·

Within this book are described over one hundred species of flowering plants which are native to the British Isles, or are well established aliens. They range from small annual herbs to tall trees, but all bear succulent fruits. The true succulent fruits are berries and simple or compound drupes; the false succulent fruits include the hips of roses, the pomes of apples and pears, and the multiple false fruits of strawberries.

The first chapter describes the various types of habitat where they will be found, indicating in each case some of the species typical of that particular habitat.

The second chapter deals briefly with the manner in which berries, drupes and pomes are formed, and their role in the spread of the species.

The chapter on toxicology contains simple definitions of the various types of poison found in the species described in this book, and advises on the basic first aid procedures to adopt in cases of suspected poisoning.

The species are arranged in the same botanical order as that used in *Flora Europaea*, and the English names given are those adopted in 'English Names of Wild Flowers'.

When dealing with complex groups, such as the brambles and roses, where there are hundreds of native species, it has been necessary to select those which are useful representatives of specific areas, and which will help the enquirer to place the identity of an unknown plant within a reasonably defined group. Further detailed descriptions of individual species are the province of specialist publications. In the text on the common whitebeams there are additional short notes on the rare and local species related to them; similar notes are appended to the text on the ivies, cranberry, bilberry and honeysuckle.

The text on each species follows the same pattern; first a detailed description of the fruit, noting the season when it ripens, then a description of the whole plant, its foliage and flowers. This is followed by notes on the ecology of the plant and its distribution within the British Isles and in Europe.

*Note* The names of British counties and vice-counties used are those current before the 1974–5 reorganization of county and regional boundaries. They have been retained for all botancial recording, which is still done according to the Watsonian county and vice-county system, otherwise there could be no continuity of historical records.

'History and Uses' describes the way in which the plant and its fruits have been used in the past and their present uses, if any. Poisonous species are indicated by p, and the texts include the toxic principles where known; the symptoms of poisoning in man and animals, and recommended treatment (i.e. first aid).

# · WHERE TO FIND BERRIES · AND FRUITS

*Hedgerows*

Hedgerows are the richest habitat for the berry hunter, and have the great advantage of accessibility to the walker and to the less active naturalist who must perforce remain car-bound.

The word 'hedge' derives from the Anglo-Saxon *gehaeg*, which is related closely to 'hay' or 'haw' – a thin strip of woodland dividing parcels of land. 'Haw' is also used for the fruit of the haw-thorn or hedge-thorn. The first Enclosure Act of 1603 was followed by a series of General Acts which continued into the 19th Century. Vast quantities of 'quickset' or live hawthorn plants were needed to provide the extensive boundary hedges. Hedges also existed on pre-enclosure boundaries, and these ancient hedges tend to be richer in the number of species which they contain.

The early 20th Century slump in agriculture brought about the encroachment of hedges on fields. In the two World Wars the demand for home produced food led to a revolution in agricultural techniques, the development of larger fields, and the wholesale removal of hedges, a process which has been especially marked since 1950.

As one of the main functions of a hedge is to provide a stock-proof boundary, hawthorn, midland hawthorn and blackthorn are common. Privet, barberry, snowberry and holly are frequent berry bearers in hedgerows, with dogwood and spindle-tree on chalky soils, and guelder-rose in damp areas. Hedgerows are the finest habitat for the hundreds of roses and brambles which grow in the British Isles; and where mature trees have been allowed to grow you may find the fruits of cherry, plum, bullace and crab apple. Hedges are useful for scrambling and climbing plants, their tubers hidden deep underneath, their fruits hanging in clusters in the autumn: plants such as black and white bryony, bittersweet and honeysuckle.

The sheltered banks below hedges provide just the right environment for wild strawberry in summer time, and for the brilliant red-berried spikes of lords-and-ladies in the autumn. Hedges near the sea, especially in the south-west of England, are full of superb blackberries and honeysuckle, with fuchsia, wild madder and, in a few places, the rare Italian lords-and-ladies.

*Woodland*

Searching for fruits in woodland entails looking upwards as well as downwards. The fruits of the whitebeams and cherries, for example, may be 5-10m above your head, while fruits of such low-growing plants as bilberry, cowberry, bramble and a host of other undershrubs, will best be found in woodland clearings where there is extra light. Each type of woodland will have its characteristic flora and hence its own list of berry-bearing plants. In the South, broadleaf woodland on chalk and limestone is the place to look for yew, wild cherry, wild service-tree and the whitebeams, and at ground level for gooseberry, the rare mezereon, spurge-laurel, Herb-Paris, lily-of-the-valley and common and angular solomon's-seal. Oakwoods will have holly and some rowan, with bilberry and butcher's-broom in the understorey. Northern birchwoods often have a carpet of bilberry, and the old pinewoods of Scotland are full of juniper, with bilberry and cowberry abundant on the damp, moss-covered ground. Alder carr is the place to find alder buckthorn and the common buckthorn, and both black and red currants are frequent there.

The scrub which so readily invades chalk grassland is a fine place to search for fruits. Hawthorn is the commonest berry bearer, with elder, dogwood, privet, and, less commonly, juniper. Both black bryony and bittersweet are common climbers, with roses and dewberry on the more open ground, and this is the habitat for deadly nightshade. The juniper bushes on chalk downland are scruffier and smaller than those of the Scottish pinewoods, but often bear a heavier crop of berries.

*Heath and Moorland*

Heath forms on light, sandy or gravel soils where the rainfall is low. Bilberry is common, and rowan also. Moorland forms over acid rock where rainfall is high, and is a feature of land above 300m, especially in the north and west. Here bilberry and crowberry are common, with cowberry, cloudberry and dwarf cornel, although the last two are seldom prolific fruit formers. Rowan is also common, particularly along the banks of streams and rivers.

*Mountains*

Crowberry and bilberry are predominant among the dwarf mountain shrubs, and in sheltered areas they can carry heavy crops of berries. In the north, bearberry and mountain crowberry are common, and bog bilberry is widespread, although it seldom bears much fruit. August and September are good times to search for the mountain berries, when the foliage of dwarf cornel turns beautiful shades of pink and purple in the more sheltered gullies where it grows, and alpine bearberry with its black fruits and leaves turning scarlet may be found on the bare shoulders of a few highland peaks.

*Limestone Pavement and Cliffs*

Limestone supports a very specialized flora, among which are some uncommon plants with succulent fruits. The limestones of the Avon and Severn valleys yield a number of rare whitebeams with a very limited distribution, and on limestone in Yorkshire and Derbyshire the main concentrations of mountain currant and stone bramble will be found. Wild cotoneaster grows only on the limestone of Great Orme's Head in North Wales, and baneberry on the limestone pavement of Yorkshire and Lancashire. Burnet rose, with its blackish purple hips, is a common feature of the limestone terraces of the Burren in western Ireland, and is equally at home on other chalk and limestone soils. Juniper is common on the limestones of northern England and central Scotland, where it fruits well.

*Bog and Fen*

These specialized habitats are the home of several berry-bearing plants. Bilberry, crowberry and cloudberry are common in the less wet parts of blanket bogs, while cranberry and small cranberry grow only in the wettest areas among the cushions of sphagnum moss. Buckthorn and alder buckthorn are widespread in fen scrub; also red and black currants, and in a few places in northern East Anglia the rare berry catchfly.

*Seaside*

Maritime habitats are not very productive of fruits and berries, but they do contain a number of specialized and interesting species. Sea-asparagus grows on a very few cliff tops and dunes in the south west and west of England and Ireland, and the same cliffs and rocks may contain wild madder, Atlantic and common ivy, Duke of Argyll's tea-plant, as well as an abundance of bramble and blackthorn. The curious dwarf forms of blackthorn and bittersweet are found only on seaside shingles, and sea-buckthorn is most abundant in coastal dunes, especially in the south and east of England. Wild fuchsia beautifies the seaside hedges in the south west and west of Britain and Ireland. The same hedges often contain honeysuckle and superb blackberries in abundance, with Italian lords-and-ladies growing in the shelter under them in a few sites along the south west coast.

*Farmland and Orchards*

Black nightshade and the two green nightshades flourish best in disturbed land, especially on the edges of crops such as potatoes and carrots. They will also be found where dung has been stacked and left; in cracked concrete at the edge of farm yards; and by buildings where the ground is rich in nitrogenous waste. Old and even derelict orchards are a favoured habitat for mistletoe.

# · THE FORMATION OF FRUITS ·

A fruit is properly defined as a dry or fleshy case developing from the wall of the ovary and surrounding the seeds, but the term is often used to describe the whole complex structure, such as a rose hip or an apple, which is constructed of associated parts of the flower including the receptacle. The receptacle is the tip of the flower stalk to which all the floral parts are attached, and it may be dome-shaped, conical, flat or cup-shaped.

Some plants develop swollen, attractively coloured or palatable fruits to attract the birds or animals which eat them. The seeds, enclosed in a hard protective coat, pass through the animal's digestive tract undamaged, to be deposited at a distance from the parent plant, thus encouraging spread of the species.

Succulent fruits are of two main kinds: the true succulent fruits which are formed only from the ovary, and the false succulent fruits which are formed by other parts of the flower.

## *True Succulent Fruits*

*Berry*

A berry contains one or more seeds enclosed within a single structure. The seeds do not have any stony, hard coat around them, and are contained in a soft, pulpy flesh formed from the ovary wall. Typical examples of plants with berries are bilberry, cranberry, gooseberry, the currants and the nightshades.

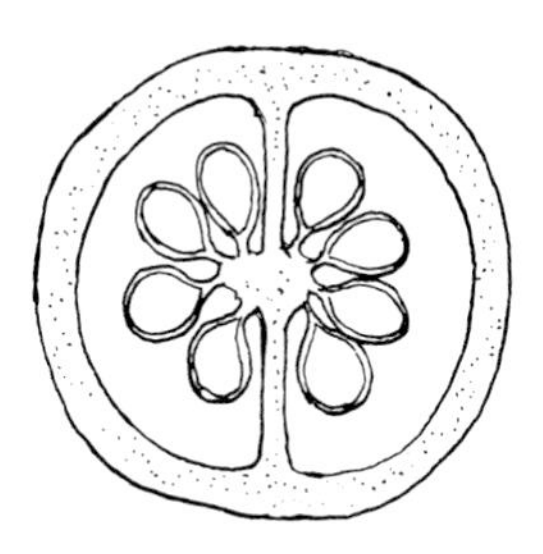

*Drupe*

A drupe is a fleshy fruit resembling a berry but containing one seed which is enclosed in a hard stony or woody case. The fleshy pericarp is formed from the ovary wall.

Cherries, sloes and plums are typical simple drupes. Blackberries and raspberries are examples of compound drupes, each fruit being composed of a number of individual drupelets arranged on a common structure formed from the receptacle.

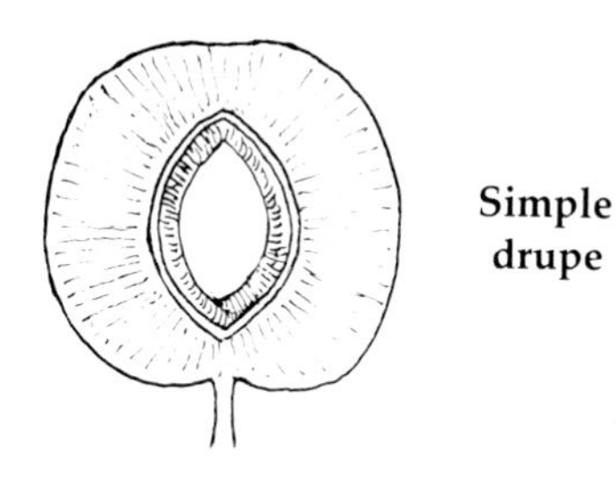

**Simple drupe**

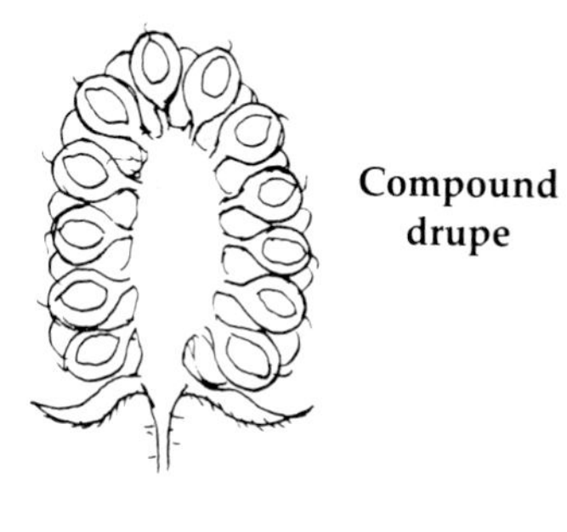

**Compound drupe**

## *False Succulent Fruits*

*Hip*

The rose hip is an example of a false succulent fruit formed by the swollen wall of the receptacle below the sepals, which in this case is urn-shaped. The ovaries lie on the inner wall of the receptacle, and the woody seeds which form are loosely attached to the wall and packed in with a mass of fine, silky hairs.

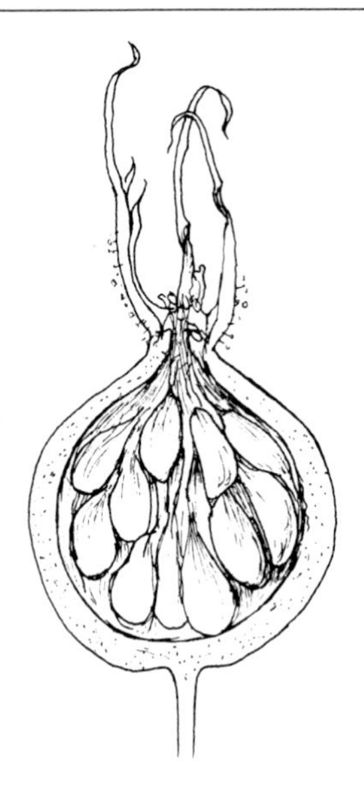

Each of the nutlets within the hip represents a true fruit, the whole mass being enclosed by the receptacle wall to form the false fruit.

*Pome*

The fruit of an apple, pear, hawthorn, cotoneaster and whitebeam is called a pome, and is very similar in its development to the rose hip. There are one to five seeds, each surrounded by a tough, but not stony, layer, the carpels in which they form being situated at the base of the concave receptacle, to which they are attached by long, sloping bases. They are also united laterally. The whole of the outer part of the fruit is then formed of the fleshy but firm cup of the receptacle.

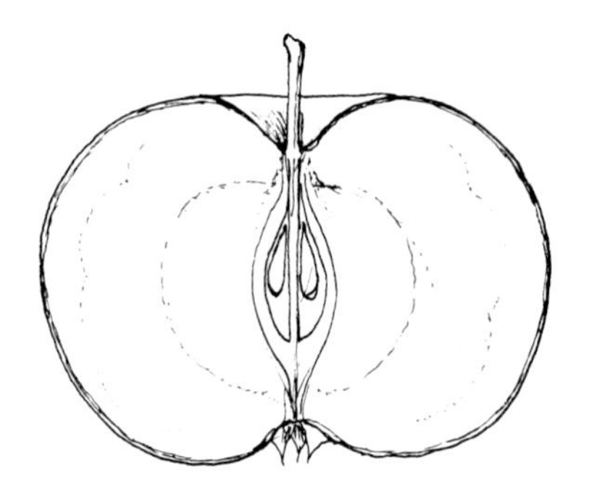

*Strawberry*

The fruit of the strawberry is a compound false fruit, the main mass being formed by the enlarged, pulpy receptacle. The true fruits are the pips set into the surface of the strawberry.

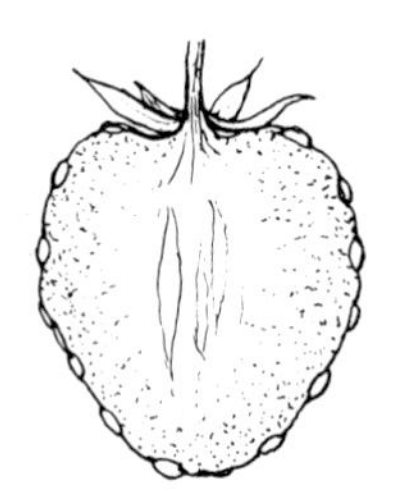

# · CAN WE EAT THEM? ·

Fruits cannot be classed simply as edible or poisonous. Nearly a third of those described in this book are either not worth eating, or possibly – but not definitely – harmful. The following lists divide the fruits into:

**A** Those which are certainly poisonous.

**B** Those which are good to eat. Others may come into this category under special circumstances.

**C** Those which may cause harm (and which are scarcely worth eating).

## *Group A*

Alder Buckthorn
Angular Solomon's-seal
Baneberry
Barberry (unripe fruits)
Bird Cherry
Bittersweet
Black Bryony
Black Nightshade
Buckthorn
Deadly Nightshade
Duke of Argyll's Tea-plant
Elderberry (unripe fruits)
Green Nightshade
Herb-Paris
Holly
Italian Lords-and-Ladies
Ivy
Juniper
Lily-of-the-valley
Lords-and-Ladies
Mezereon
Mistletoe
Snowberry
Solomon's-seal
Spindle-tree
Spurge-laurel
Tutsan
White Bryony
Whorled Solomon's-seal
Wild Privet
Yew

## *Group B*

Barberry (ripe fruits only)
Bilberry
Blackthorn (sloe)
Black Currant
Bog Bilberry
Bramble (blackberry)
Bullace
Cherry Plum
Cloudberry
Cowberry
Crab Apple
Cranberry
Dewberry
Dwarf Cherry
Dwarf Cornel
Elderberry
Garden Strawberry
Gooseberry
Juneberry
Medlar (when bletted)
Mountain Crowberry
Mountain Currant
Oregon-grape (ripe fruits only)
Raspberry
Red Currant
Rose hips (as conserve or syrup)
Rowan
Sea-buckthorn
Stone Bramble
Strawberry-tree
Wild Pear (if fully ripe)
Wild Plum
Wild Service-tree (when bletted)
Wild Strawberry

## *Group C*

Alpine Bearberry
Bearberry
Berry Catchfly
Butcher's-broom
Common Whitebeam
Crowberry
Dogwood
Dwarf Elder
Fly Honeysuckle
Fuchsia
Guelder-rose
Hawthorn
Himalayan Cotoneaster
Honeysuckle
May Lily
Midland Hawthorn
Plymouth Pear
Sea Asparagus
Small Cranberry
Small-leaved Cotoneaster
Swedish Whitebeam
Wall Cotoneaster
Wayfaring-tree
Wild Asparagus
Wild Cotoneaster
Wild Madder

These lists refer *only* to the fruits, and for further information on the edibility of other parts of the plants, the full text must be consulted.

*Never* eat wild fruits unless you are completely certain that they are safe, and that you have identified them correctly.

# · PLANT POISONS ·

It is wise to treat all unknown fruits as potentially poisonous, and not to eat wild fruits unless you know for certain that it is safe to do so.

The presence of a poisonous plant at a poisoning incident should not be taken as evidence that it is the cause of the problem. I have seen a magnificent stand of the highly toxic thorn-apple (*Datura stramonium*), growing in a cattle yard where the cattle were suffering from poisoning which ultimately proved to have been caused by their licking old lead paint. The thorn-apple had grown there for more than ten years, but they had always ignored it.

The toxic fruit and the sufferer might be brought together by artificial circumstances, where normally they would not meet. Christmas decoration with holly and mistletoe has resulted in poisoning both of children and of household pets who have eaten the berries. Natural circumstances, such as heavy snowfall, will break branches off yew trees, so that grazing animals gain access to the poisonous foliage which had previously been out of reach.

The response of individuals to a particular poison will vary, depending upon the age and state of health of the patient: the very young and the old being particularly susceptible. Intercurrent disease involving the liver or kidneys may prevent the body detoxicating a poison which would normally cause little harm. The presence of other food in the digestive tract may dilute the effects of poisonous material or delay the onset of symptoms. Whatever the condition of the person or animal at risk, all cases of possible poisoning should be treated promptly, and medical or veterinary advice should be urgently sought.

Among animals there is considerable species variation in the response to poisonous plants. Birds will eat with relish the berries of mezereon (*Daphne mezereum*) which would blister the human mouth, and rabbits can eat deadly nightshade (*Atropa belladonna*) without harm, since the toxic principle atropine is destroyed by atropinesterase produced in the liver of the rabbit.

The amount of the poisonous principle may vary in different parts of the plant, and plants of the same species grown under different conditions of soil or climate may be innocuous or highly poisonous.

Most poisonous plants are unpalatable, so that animal poisoning is not common, but occasionally animals will develop depraved appetites and go to extraordinary lengths to get at supplies.

## *Classification of Plant Poisons*

Plant poisons are often highly complex in their chemical structure. A further complication arises from the fact that some species contain chemicals from more than one group, (e.g. an alkaloid and a glycoside). The following simplified list includes most of the types of compound described in this book.

Alkaloids
Glycosides – cyanogenic
– cardiac
– coumarins and furanocoumarins
– saponins
– (aglycones)
Toxic amino-acids
Essential or volatile oils
Oxalates

### *Alkaloids*

Alkaloids are complex organic nitrogenous compounds produced by plants. They consist of many different types and are often bitter-tasting. Their function within the plant is not known, but they can have a profound effect upon the central and peripheral nervous system of animals and man. A single substance may cause a variety of different physiological responses. The following list includes some typical alkaloids and their plants of origin.

*Steroidal Alkaloids*
Solanine from bittersweet
Protoveratrine from some of the Liliaceae.

*Indole Alkaloids*
Strychnine from nux vomica. Few European plants have alkaloids of this type.

*Isoquinoline Alkaloids*
Morphine from opium poppy.
Emetine from ipecacuanha.

*Quinolizidine Alkaloids*
Cytisine and sparteine from lupin and laburnum.

*Terpenoid Alkaloids*
Aconitine from delphinium.
Unidentified terpenes from snowberry.

*Tropane Alkaloids*
Hyoscyamine from thorn-apple and deadly nightshade.
Atropine from deadly nightshade.
Hyoscine from henbane.

*Other Alkaloids*
Colchicine from meadow saffron.
Taxine from yew.

*Glycosides*
Glycosides are a large group of organic substances consisting of one or more monosaccharides (simple sugars), combined with a non-sugar called an aglycone. They are normally non-toxic within undamaged plant tissue, but if the tissue is macerated or digested the action of water, enzymes or acids will split the glycoside into its constituent sugar and aglycone. The aglycones are pharmacologically active, and are classified according to the type of activity which they demonstrate.

Cyanogenic glycosides are split by enzymes within the plant itself which only come into contact with the glycoside if the plant tissue is damaged. Hydrocyanic acid is then released, which is profoundly damaging to the process of cell respiration in the poisoned animal, particularly affecting cells with a high metabolic rate, such as those of the heart muscle and central nervous system.

Cardiac glycosides have a specific action on the heart muscle and its nerve supply, causing more powerful contractions and a slowing of the heart rate.

Coumarin glycosides are highly irritant, causing blistering of the mouth and digestive tract.

Purgative glycosides on breakdown yield the strongly purgative anthraquinone.

Saponins are glycosides of high molecular weight and bitter taste which produce a frothing effect upon water. They are not normally of great toxicity to mammals unless they enter the bloodstream, when they will destroy red blood-cells (haemolysis).

The following list contains some typical glycosides and their plants of origin.

*Cyanogenic Glycosides*
Amygdalin from bird cherry.
Amygdalin from seeds of *prunus* and *malus.*
Unidentified glycoside in Lords-and-Ladies.

*Cardiac Glycosides*
Taxine from yew.
Euonymin from spindle-tree.
Convallamarin from lily-of-the-valley.

*Coumarin Glycosides*
Hypericin from St John's-wort.
Daphnin from mezereon and spurge-laurel.

*Protoanemonin Glycosides*
Protoanemonin from baneberry.

*Purgative Glycosides*
Rhamnosterin and rhamnicoside from buckthorn.
Frangulin and glucofrangulin from alder buckthorn.

*Saponin Glycosides*
Hederasaponins A and B from ivy.
Dulcamaric acid from bittersweet.
Paristyphnin from Herb-Paris.
Arin from Lords-and-Ladies.

*Aglycones*
Solanidine from black nightshade.
Diosgenin from Solomon's-seal.

---

*Toxic Aminoacids*
Examples include the numerous viscotoxins which have been isolated from mistletoe.

*Essential Oils*
Examples include the terpene hydrocarbons limonene, and $\alpha$ and $\beta$ pinene from juniper.

*Oxalates*
Another group of poisonous substances are salts called oxalates, which are present in the Arums and

also in the berries of black bryony. Tiny crystals of insoluble calcium oxalate are capable of producing intense irritation and burning of the mouth and gastrointestinal tract. The soluble oxalates cause severe damage to the kidneys. In the bloodstream they combine with calcium ions to produce calcium oxalate and a drop in the available blood calcium. Cows poisoned with arum thus appear similar to those suffering from milk fever – the physiological drop in blood calcium which causes recumbency and coma in high-yielding cows at calving time.

## *Diagnosis, Assessment and Treatment*

Diagnosis in cases of plant poisoning is not easy, and will depend upon the symptoms shown and the circumstantial evidence of the incident. It is as well to remember that in cases of poisoning not all the possible symptoms will be produced every time, and post-mortem lesions are rarely characteristic of a particular poison. Always keep any suspect material and give it to the doctor or veterinary surgeon examining the case; keep both the suspect plant material and anything vomited or passed by the patient.

In all cases of acute poisoning prompt treatment is essential. It should aim (a) to prevent further absorbtion of the toxic material, (b) to assist removal of the substance from the stomach, (c) to provide monitoring and support of vital functions and, (d), where applicable, to include the administration of a specific antidote.

The general principles of management following the ingestion of poisonous plants found in Great Britain are, however, relatively straightforward.

In describing the treatment for poisoning in animals, where a specific remedy is possible, this is mentioned in the text. However, cases of animal poisoning with berries are relatively rare, and for some of the species mentioned here no precedent for particular therapy exists. In these cases the basic guidelines as set out for human treatment can be followed, but in view of the variable response which different animal species can show to the drugs advised, veterinary supervision is most desirable.

### *Treatment of the Conscious Patient*

If necessary the stomach should be emptied to prevent further absorbtion of the plant material. In the conscious patient the administration of an emetic (usually syrup of Ipecacuanha BP) will cause prompt vomiting. Where this is not available, or where there is likely to be an unavoidable delay before medical treatment is available, emesis may be induced by *gently* tickling the back of the throat with clean fingers or with the end of a spoon handle. The patient should be kept in a head-down position, or, in the case of a child, held over the knee with the head lower than the hips. **SALT WATER EMETICS SHOULD NOT BE USED AS THEY ARE NOT ONLY INEFFECTIVE BUT ALSO DANGEROUS.**

In animals with a simple stomach, vomiting can be induced by administering a large crystal or two of washing soda (sodium sulphate). This works well in dogs and cats, but not in ruminants with their complex stomachs or in horses, which cannot vomit.

The stomach may also be emptied by gastric lavage, which should be performed under medical supervision. In cases of serious poisoning in adults, or where a large quantity has been ingested, or where the patient is unconscious, a stomach washout is advisable.

If a highly irritant or corrosive substance has been ingested, vomiting should *not* be induced. Oral demulcents such as milk, beaten egg whites or vegetable oil should be given to dilute the substance and to soothe the gut.

Adsorbants such as activated charcoal, 'Medicoal' or 'Carbomix', can be given orally. 10g is usually sufficient for most plant poisons in Great Britain. The 'Universal Antidote' has been proved to be ineffective and is no longer used.

Purgatives or laxatives are sometimes given to increase the speed of evacuation of the ingested material. These agents should be used with caution, as they may increase fluid loss and cause electrolyte imbalance and dehydration.

## *Treatment of the Unconscious Patient*

The unconscious patient should be checked to ensure that he or she is still breathing, that the airway is clear, and that the pulse is palpable. If not, the patient should be resuscitated immediately. The patient should be kept in the 'Recovery Position', (semi-prone with the head turned to the left side to prevent aspiration of vomitus and to keep the airway patent), whilst emergency help is sought.

In cases of severe poisoning, vital functions should be monitored, and ventilatory or cardiovascular support given as necessary.

Should a stomach wash-out be indicated, it should be performed with a cuffed endotracheal tube in place to stop material getting into the airway. Problems can arise from berries blocking the tube inlet. 10 g of activated charcoal may be left in the stomach following lavage.

The majority of cases involving the ingestion of berries from British plant species are resolved within twenty-four hours without permanent damage.

Specialist services are available on a twenty-four hour basis to advise doctors dealing with cases of poisoning, and data are correlated through the National Poisons Information Service (NPIS).

The Author would be glad to receive any data on confirmed cases of poisoning by berries or fruits, either in man or in animals, so that the information can be passed on to the NPIS and used to increase our knowledge of such cases and their treatment.

# · SPECIES LIST ·

## p = poisonous

|  |  | |  | | |
|---|---|---|---|---|---|
| 1 | Juniper | p | *Juniperus communis* L. | | |
| 2 | Yew | p | *Taxus baccata* L. | | |
| 3 | Mistletoe | p | *Viscum album* L. | | |
| 4 | Berry Catchfly | ? p | *Cucubalus baccifer* L. | | |
| 5 | Baneberry | p | *Actaea spicata* L. | | |
| 6 | Barberry | p | *Berberis vulgaris* L. | | |
| 7 | Oregon-grape | p | *Mahonia aquifolium* (Pursh) Nuttall | | |
| 8 | Red Currant | | *Ribes rubrum* L. | | |
| 9 | Black Currant | | *Ribes nigrum* L. | | |
| 10 | Gooseberry | | *Ribes uva-crispa* L. | | |
| 11 | Mountain Currant | | *Ribes alpinum* L. | | |
| 12 | Cloudberry | | *Rubus chamaemorus* L. | | |
| 13 | Stone Bramble | | *Rubus saxatilis* L. | | |
| 14 | Raspberry | | *Rubus idaeus* L. | | |
| 15 | Bramble | | *Rubus fructicosus* L. | Subsection | *Suberecti* |
| | | | | Subsection | *Corylifolia* |
| | | | | Subsection | *Silvatici* |
| | | | | Subsection | *Discolores* |
| | | | | Subsection | *Sprengeliani* |
| | | | | Subsection | *Appendiculati* |
| | | | | Subsection | *Glandulosi* |
| | | | (inc. *R.spectabilis* and *R.procerus*) | | |
| 16 | Dewberry | | *Rubus caesius* L. | | |
| 17 | Field Rose | | *Rosa arvensis* Hudson | | |
| 18 | Burnet Rose | | *Rosa pimpinellifolia* L. | | |
| 19 | Stylous Rose | | *Rosa stylosa* Desvaux | | |
| 20 | Dog Roses | | *Rosa coriifolia* Fries | | |
| | | | *Rosa afzeliana* Desvaux | | |
| | | | *Rosa canina* L. | | |
| | | | *Rosa obtusifolia* Desvaux | | |
| 21 | Downy Roses | | *Rosa tomentosa* Smith | | |
| | | | *Rosa sherardii* Davies | | |
| | | | *Rosa mollis* Smith | | |
| 22 | Sweet Briars | | *Rosa rubiginosa* L. | | |
| | | | *Rosa micrantha* Borrer | | |

| | | | |
|---|---|---|---|
| 23 | Wild Strawberry | | *Fragaria vesca* L. |
| 24 | Garden Strawberry | | *Fragaria* x *ananassa* Duchesne |
| 25 | Plymouth Pear | | *Pyrus cordata* Desvaux |
| 26 | Wild Pear | | *Pyrus pyraster* Burgsdorff |
| 27 | Crab Apple | | *Malus sylvestris* Miller |
| 28 | Rowan | ? p | *Sorbus aucuparia* L. |
| 29 | Wild Service-tree | | *Sorbus torminalis* (L.) Crantz (inc. *S. domestica*) |
| 30 | Common Whitebeam | | *Sorbus aria* (L.) Crantz (inc. *S.rupicola, S.porrigentiformis, S.hibernica, S.wilmottiana, S.eminens, S.vexans, S.leptophylla, S.lancastriensis*) |
| 31 | Swedish Whitebeam | | *Sorbus intermedia* (Ehrhart) Persoon (inc. *S.anglica, S.minima, S.leyana, S.arranensis, S.pseudo-fennica, S.devoniensis* |
| 32 | Broad-leaved Whitebeam | | *Sorbus latifolia* (Lamarck) Persoon (inc. *S.bristoliensis, S.subcuneata, S.decipiens*) |
| 33 | Juneberry | | *Almelanchier lamarckii* F.Schroeder (inc. *A.spicata, A.confusa* & *A.ovalis*) |
| 34 | Wall Cotoneaster | | *Cotoneaster horizontalis* Decaisne |
| 35 | Himalayan Cotoneaster | | *Cotoneaster simonsii* Baker |
| 36 | Wild Cotoneaster | | *Cotoneaster integerrimus* Medicus |
| 37 | Small-leaved Cotoneaster | | *Cotoneaster microphyllus* Wallich ex Lindley |
| 38 | Medlar | | *Mespilus germanica* L. |
| 39 | Midland Hawthorn | | *Crataegus laevigata* (Poiret) DC |
| 40 | Hawthorn | | *Crataegus monogyna* Jacquin |
| 41 | Cherry Plum | | *Prunus cerasifera* Ehrhart |
| 42 | Blackthorn | | *Prunus spinosa* L. |
| 43 | Wild Plum | | *Prunus domestica* L. |
| 44 | Bullace | | *Prunus domestica* subsp. *insititia* (L.) C.K. Schneider |
| 45 | Wild Cherry | | *Prunus avium* (L.) L. |
| 46 | Dwarf Cherry | | *Prunus cerasus* L. |
| 47 | Bird Cherry | p | *Prunus padus* L. |
| 48 | Holly | p | *Ilex aquifolium* L. |
| 49 | Spindle-tree | p | *Euonymus europaeus* L. |
| 50 | Buckthorn | p | *Rhamnus catharticus* L. |
| 51 | Alder Buckthorn | p | *Frangula alnus* Miller |
| 52 | Mezereon | p | *Daphne mezereum* L. |
| 53 | Spurge-laurel | p | *Daphne laureola* L. |
| 54 | Sea-buckthorn | | *Hippophaë rhamnoides* L. |
| 55 | Tutsan | ? p | *Hypericum androsaemum* L. |
| 56 | White Bryony | p | *Bryonia cretica* L. subsp. *dioica* (Jacquin) Tutin |

| | | | |
|---|---|---|---|
| 57 | Fuchsia | | *Fuchsia magellanica* Lamarck sensu stricto (inc. var Riccartoni) |
| 58 | Dogwood | | *Cornus sanguinea* L. |
| 59 | Dwarf Cornel | | *Cornus suecica* L. |
| 60 | Ivy | p | *Hedera helix* L. |
| | | p | *Hedera hibernica* (Kirchner) Bean (inc. *H.colchis* and *H.algeriensis*) |
| 61 | Strawberry-tree | | *Arbutus unedo* L. |
| 62 | Bearberry | | *Arctostaphylos uva-ursi* (L.) Sprengel |
| 63 | Alpine Bearberry | | *Arctostaphylos alpinus* (L.) Sprengel |
| 64 | Cranberry | | *Vaccinium oxycoccos* L. (inc. *V.macrocarpon* Aiton) |
| 65 | Small Cranberry | | *Vaccinium microcarpum* (Turczaninow ex Ruprecht) Schmalhausen |
| 66 | Cowberry | | *Vaccinium vitis-idaea* L. |
| 67 | Bog Bilberry | | *Vaccinium uliginosum* L. |
| 68 | Bilberry | | *Vaccinium myrtillus* L. (inc. *V.corymbosum* L.) |
| 69 | Crowberry | | *Empetrum nigrum* L. |
| 70 | Mountain Crowberry | | *Empetrum hermaphroditum* Hagerup |
| 71 | Wild Privet | p | *Ligustrum vulgare* L. |
| 72 | Duke of Argyll's Tea-plant | p | *Lycium barbarum* L. (inc. *L.chinense* Miller) |
| 73 | Deadly Nightshade | p | *Atropa belladonna* L. |
| 74 | Black Nightshade | p | *Solanum nigrum* L. (inc. ssp.*schultesii* (Opiz) Wessely |
| 75 | Green Nightshade | p | *Solanum nitidibaccatum* Bitter |
| | | p | *Solanum sarrachoides* Sendtner |
| 76 | Bittersweet | p | *Solanum dulcamara* L. |
| 77 | Wild Madder | | *Rubia peregrina* L. |
| 78 | Dwarf Elder | | *Sambucus ebulus* L. |
| 79 | Elder | p | *Sambucus nigra* L. |
| 80 | Guelder-rose | | *Viburnum opulus* L. |
| 81 | Wayfaring-tree | | *Viburnum lantana* L. |
| 82 | Snowberry | p | *Symphoricarpos albus* (L.) S.F.Blake |
| 83 | Fly Honeysuckle | | *Lonicera xylosteum* L. |
| 84 | Honeysuckle | | *Lonicera periclymenum* L. (inc. L. caprifolia) |
| 85 | Lily-of-the-valley | p | *Convallaria majalis* L. |
| 86 | May Lily | | *Maianthemum bifolium* (L.) F.W.Schmidt |
| 87 | Whorled Solomon's-seal | p | *Polygonatum verticillatum* (L.) Allioni |
| 88 | Solomon's-seal | p | *Polygonatum multiflorum* (L.) Allioni |

| | | | |
|---|---|---|---|
| 89 | Angular Solomon's-seal | p | *Polygonatum odoratum* (Miller) Druce |
| 90 | Herb-Paris | p | *Paris quadrifolia* L. |
| 91 | Wild Asparagus | | *Asparagus officinalis* L.subp.*officinalis* |
| 92 | Sea Asparagus | | *Asparagus officinalis* L.subsp.*prostratus* (Dumortier) Corbière |
| 93 | Butcher's-broom | | *Ruscus aculeatus* L. |
| 94 | Black Bryony | p | *Tamus communis* L. |
| 95 | Italian Lords-and-Ladies | p | *Arum italicum* subsp.*neglectum* (Townsend) Prime |
| 96 | Lords-and-Ladies | p | *Arum maculatum* L.• |

# · JUNIPER ·

***Juniperus communis* L.**

POISONOUS

*Fruit*

The fruits of juniper are formed of fleshy cone scales, which coalesce to produce a round, berry-like structure 6-9mm in diameter. Each fruit is solitary, borne on a very short stem in the axils of the needle-shaped leaves, but the fruits are often clustered together. In the first year they are green with a blue sheen, ripening through brown to purple black in the second year, so that a mixture of different coloured fruits can be seen at the same time.

The skin of the fruit is shiny, tough, and leathery, the apex bearing three scars like small teeth.

There is little flesh in the fruit, each of which contains two or three (rarely six) triquetrous, brown seeds 4mm long, sharply pointed and with a rough, pitted base. The flavour of the fruit is strongly aromatic, not unpleasant if you like the taste of gin, and will remain for hours in the mouth when even a single berry is chewed. Ripe juniper berries are best sought in October.

*Leaf and Flower*

Juniper is an evergreen shrub 2-10m high, spreading and bushy in shape, or on occasion narrowly conical. It is noticeable that strongly growing bushes with graceful, arched foliage, have few berries, which are more common on stunted, stubby bushes. The leaves are needle-like, 10-20mm long, carried in whorls of three, those of the normal form being narrow and sharply pointed, with a glaucous band down the centre of the concave undersurface. The yellow male and female flowers are produced from mid-May to late July, and are usually carried on separate plants, male flowers in cones 8mm long with five or six whorls of scales, and female flowers in solitary cones 2mm wide, bearing three to eight fleshy scales which coalesce to form the berry. Flowering can be sparse or infrequent.

Subspecies *nana* Syme is a prostrate, dwarf form which grows in montane areas and on the coast of north west Scotland, Wales, the Lake District and Ireland. The needles are greyer, broader and blunter with a white band on the lower surface, and are close set so as to form a dense mat of growth. The berries are smaller, longer than broad, and fewer in number; the seeds are marked with vertical ridges. It is possible that this form was representative of the low growing shrubs of the Full Glacial flora.

*Ecology and Distribution*

Juniper is widely distributed but decreasing on chalk downs in the south of England, where it

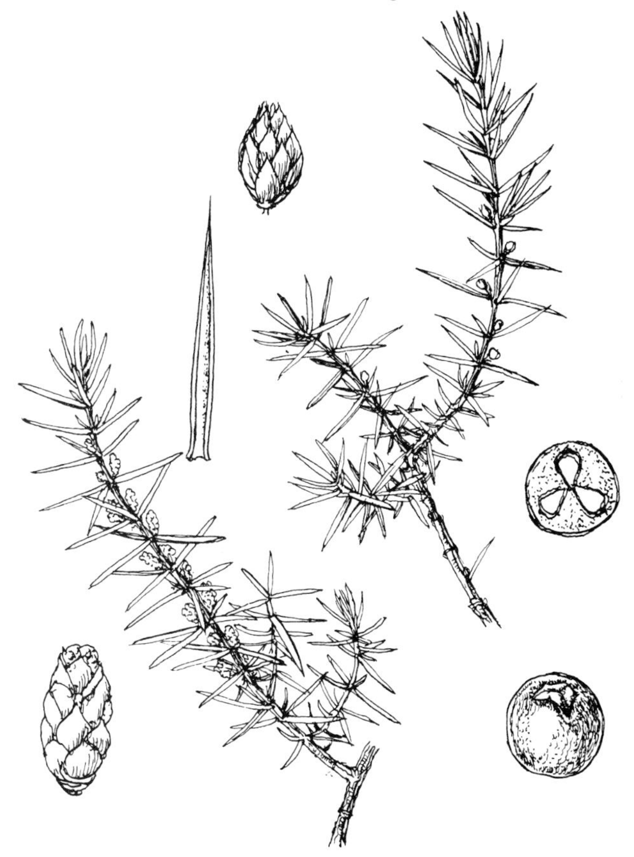

forms a special type of scrub, especially in Berkshire and Wiltshire. In the north it grows on heaths and moors, and can be abundant in the remnants of old pine forest along the flushed margins of damper degenerating areas, as in Rothiemurchus Forest, Inverness-shire and Tynron Juniper Forest, Dumfries.

In the north of England and southern Scotland it was at one time abundant, crowning the knolls on semi-moorland, and it still grows strongly on sheltered valley slopes in many places such as those north of Tomintoul in Cairngorm. In the north west it can be found right down to sea level on rocky coasts such as those of Ardnamurchan, and on sea cliffs near Aberdeen. It is scarce in the Midlands, east England and most of Wales. In Ireland it is mainly on the north and west coasts, the dwarf form occuring on Slieve League in Donegal.

In montane areas it grows most happily in gorges and gullies where erosion produces a constant source of basic elements. Its growth is extremely slow and its trunk girth is very unreliable as an estimate of age. It may increase by as little as 0.03 mm in a year, or as much as 1.97 mm. The chalk down form lives for about one hundred years in southern England, while in the north individuals can survive as long as two hundred years. In many areas it is dying out. The fertility of old bushes is much reduced, and is associated with a high rate of attack on the berries by fungi and insects, especially *Eriophyiid* mites. An investigation at Kingley Vale in October 1972 showed that eighteen out of twenty-two berries from old bushes had been destroyed. Rabbit grazing can damage young plants severely and prevent the regeneration of a population, although summer grazing to reduce competitive herbage, and some ground disturbance, may be necessary for young plants to establish. Clonal growth can occur in the prostrate form in the north and on mountains, but does not occur on chalk soils.

Juniper grows throughout Europe, but mainly on mountains in the south.

*History and Uses*

The berries of juniper have been used medicinally for many centuries as a carminative, vermifuge and diuretic. Culpeper noted that 'Indeed there is no better remedy for wind in any part of the body, or the colic, than the chymical oil drawn from the berries . . . The berries stay all fluxes, help the hemorrhoids or piles, and kill worms in children.' Gent (1681) wrote that 'Juniper berries are good against Poison, Plague, the biting of any venomous Creature, provokes Urine, is good for the Dropsey, strengthens the Stomack, expels Wind; they are good for the Cough, Shortness of Breath, Consumptions, Pains in the Belly, Ruptures, Cramps, and strengthens all the Members of the Body'. Deakin (1871) reported that 'The berries are used for flavouring Hollands and English gin. When the berries are boiled with water, they give out a quantity of sugar, which according to Linnaeus, when fermented, forms a common drink of the Lapps in Sweden.' Although nowadays this is not used commercially, people in Sweden do make a table beer at Christmas time, and you can purchase 'Aether-

**Juniper**

oleum juniperi' for the purpose of flavouring. Juniper exudes a white gum which in days gone by was sold as 'Sandarach', and in the powdered form as 'Pounce'.

Juniper berries are excellent in marinades and sauces or rubbed on to the surface of roast meats. They are an ingredient of the pickle for sauerkraut, and are very tasty crushed and mixed into a stuffing for pork. They are still used to flavour gin, and according to Messrs Tanqueray Gordon and Co Ltd, the berries that they use come from the Mediterranean, chiefly from Italy and Yugoslavia.

*Toxicology*

The needles, shoots, and particularly the berries, contain a high proportion of chemical substances; 0.5-2.0% essential oils (mainly the terpene hydrocarbons α-pinene, β-pinene and limonene), a bitter alkaloid-like substance called juniperine, sesquiterpenes (α-carophyllene, cardinene and elemene), junionone, 10% resin, 30-33% sugars and organic acids. These give the ripe fruits their characteristic odour and bitter, burning taste. Juniper berries are diuretic, and as they invoke gastric secretions are used as a stimulant and carminative. An alcoholic extract of the fruits has been used for rheumatic conditions.

**Symptoms** Overdosage of juniper berries, and more particularly of the oil distilled from the ripe fruits, may cause convulsions, as well as inflaming the alimentary and urinary tracts simultaneously. Their use should be avoided during pregnancy, and they should not be given to any patient with renal disease.

The toxic or therapeutic action of juniper appears to be dependant upon the conditions under which it is grown. There is one record of poisoning in a goat, which developed haematuria as a result of eating juniper.

**Treatment** Supportive therapy should be given where indicated, and activated charcoal administered by mouth. If juniper oil has been swallowed, medical care should be sought immediately, since it can cause life-threatening convulsions, especially in children. No cases of human poisoning have been recorded in the British Isles.

# · *YEW* ·

***Taxus baccata* L.**

POISONOUS

*Fruit*

The ripe fruit of yew is easily recognized, consisting of a smooth, brown ovoid seed 5-6mm long, enclosed in a pink fleshy structure like an open-mouthed bell, called the aril. In early summer as the fruits form, both the single seed and the aril are green, the aril at this stage forming a narrow ring from which the seed projects, like an acorn in its cup. The seed darkens to a purplish colour, while the aril enlarges and begins to enclose it. Finally the aril swells and ripens, becoming pink and fleshy, hiding the brownish-purple seed which can be seen through the open end of the bell.

The fruit, 11 × 8mm, is ripe from late August throughout October. The aril is fleshy, mucilagenous and sweet to the taste, and is eaten by a wide variety of birds, the poisonous seed passing undigested through the intestinal tract.

*Leaf and Flower*

Yew is a large evergreen tree up to 20m high, with red-brown bark and usually more than one trunk to the tree. The leaves are flat needles 10-30mm long, dark green and glossy, with two pale green bands beneath, set in two lateral pinnate series which are spirally arranged. The flowers open from February until late April, male and female flowers on separate trees. The male flowers are formed of six to fourteen peltate anthers like a little yellow mulberry 5mm in diameter, carried in the leaf axils. The female flowers are composed of scales, forming an ovoid mass 2mm long.

## *The Irish Yew*

*T.baccata* 'Fastigiata' forms a sombre, dark erect tree, with confined, spire-like branches. It is widely

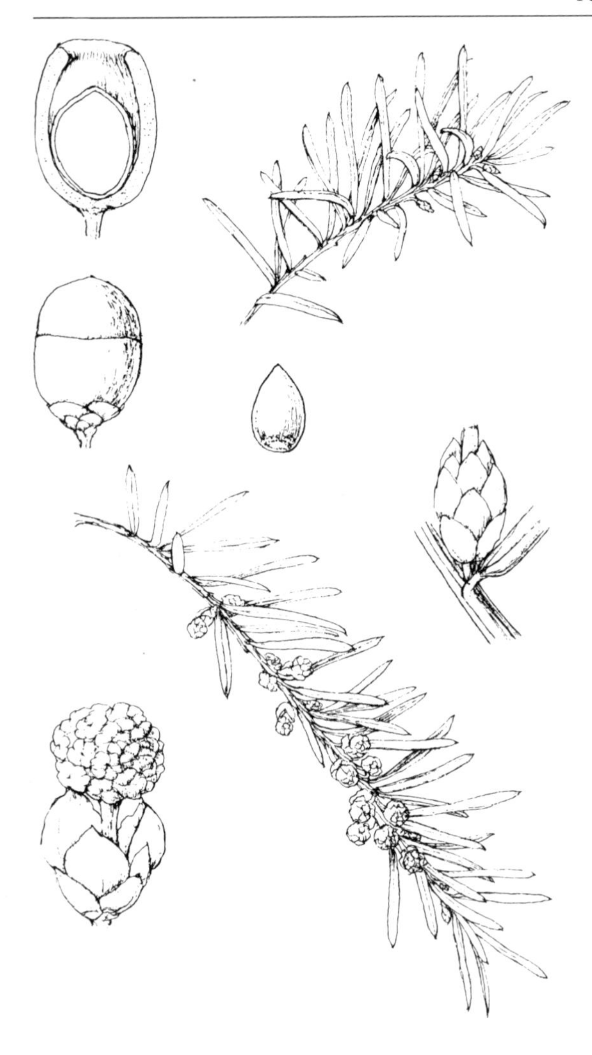

used as a hedging plant and in topiary. It is a female clone, but has been recorded carrying male strobili, although yew is normally dioecious. It was first found on the slopes of Cuilcagh Mountain, Co. Fermanagh in about 1740 by Mr George Willis, a tenant of the Earl of Enniskillen. Irish Yew was introduced for commercial purposes by George Cunningham of Liverpool, who received cuttings from Lord Enniskillen about 1780. The mother plant still survives at Florencecourt, Co Fermanagh.

*Ecology and Distribution*
Yew is rather common on the chalk and limestone of south and central England and is a characteristic component of the 'hangers' of the North Downs in Kent and Surrey, and in the Chilterns. It is a frequent member of beechwood and chalk scrub, and there are splendid stands of ancient yews in a few sites such as Kingley Vale Reserve in West Sussex. Yew is native to east Wales, and yew trees are a most impressive sight growing out of clefts in the vertical limestone cliffs of Yew Cogar Scar at Arnecliffe in Littondale, Yorkshire. Elsewhere they are likely to have been planted. They can live to a very great age.

In northern Europe, yew is to be found only at low altitudes, while in the Mediterranean it grows only on the higher mountains.

*History and Uses*
Yew trees are found in churchyards in many parts of England, where they are at least relatively safe from the attentions of children and grazing animals. It is said that they were planted there to provide a supply of wood for longbows, but longbows needed to be made of long, straight, knot-free sections, which are uncommon in our churchyard yews. Although some bows could well have been cut from English trees, it is said that the best bow-staves were imported from Spain.

*Toxicology*
Apart from the edible aril, all parts of the yew are extremely poisonous, especially the leaves and seeds. The toxic principles are resistant to drying and storage. The most important toxic ingredients are the alkaloids taxine A and B, which are non-irritant and rapidly absorbed from the intestine, leading to symptoms of poisoning within thirty minutes. Yew also contains an irritant volatile oil which causes colic; a cardiac glycoside taxiphyllin; ephedrine; and five to eight unidentified compounds. The chief effect is upon the heart and blood vessels, sudden death resulting from cardiac arrest. This can occur as soon as five minutes after ingestion, and there are instances of cattle having been found dead with sprigs of yew still in their mouths.

**Symptoms** In addition to its cardiotoxic effects, yew also affects the central nervous system and acts as a local irritant to the gut. Symptoms in man commence with dizziness, a dry mouth, dilated pupils and lethargy. Vomiting, diarrhoea and abdominal pain follow rapidly, with slowing of the heart rate, depression of cardiac conduction and a fall in blood pressure. There is difficulty in breathing; trembling and convulsions; and in severe cases coma, followed by respiratory and cardiac failure. Animals lasting for any length of time show symptoms of diarrhoea and vomiting, colic, delirium and convulsions, with a markedly slowed heart rate.

Grazing animals will eat yew, especially clippings or branches broken off by snow, when grazing is sparse. Horses and pigs are very sensitive to poisoning by yew. The lethal dose for a horse is 100-200g; for a pig 75g; for a cow 500g.

In the New Forest there is some evidence that resident ponies and cattle have developed a tolerance to yew, and they have been seen browsing it in quantity with apparent impunity.

**Treatment** Vomiting should be induced or gastric lavage performed as soon as possible, as the alkaloids are very rapidly absorbed from the gastrointestinal tract. There is no antidote to the alkaloids, so that treatment must be symptomatic and simply supportive. The heart should be monitored by ECG where possible.

More than thirty cases have been recorded by NPIS; where adults have been involved, poisoning has often been intentional and symptoms more pronounced. No fatal cases have been recorded in the last decade.

In veterinary practice poisoning, especially of ruminants, is fairly frequent and usually fatal. Where cattle are known to have eaten yew the best course of action is to perform a ruminotomy as speedily as possibly, and literally to bale out the ruminal contents with the offending yew leaves.

Seventeen species of birds have been recorded eating yew berries from August to January, the most common being song thrushes, and less frequently mistle thrushes, blackbirds and redwing. There is an isolated record of wrens eating the berries in September. The seed coat is resistant to the birds' digestive enzymes, so if it is eaten whole the berry causes no poisoning.

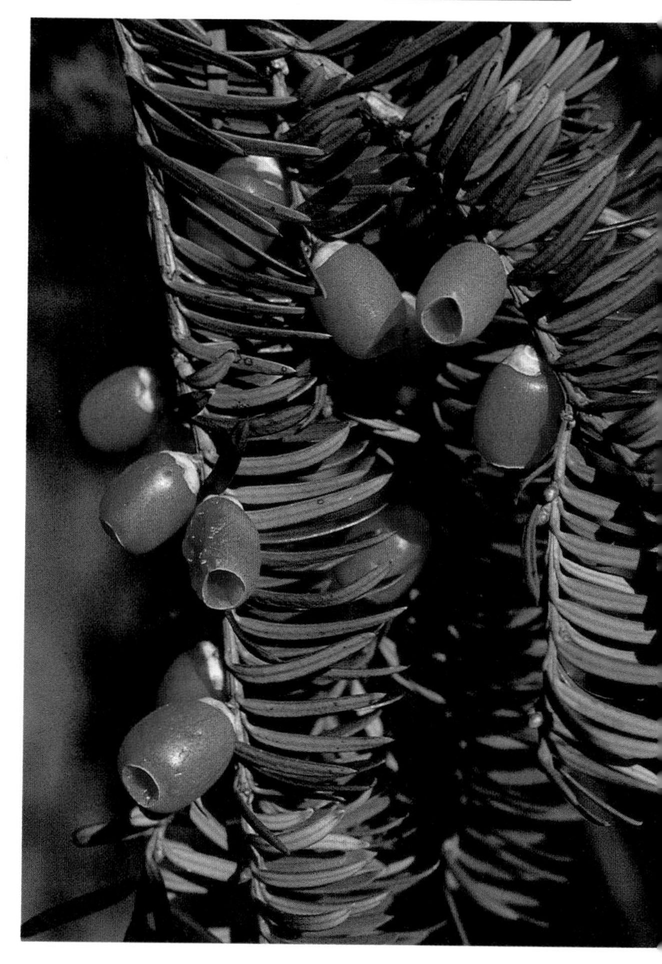

**Yew**

# · *MISTLETOE* ·

***Viscum album*** **L.**

POISONOUS

*Fruit*

The berries of mistletoe are familiar as part of traditional Christmas decorations and their mystic associations are much older than Christianity. Mistletoe berries are borne in groups of three to five on very short pedicels, the spherical fruit being greenish-white, 6-10mm in diameter, and translucent when ripe. The fibres in the berry coat give the ripe berry a veined appearance, rather like that of a gooseberry, and the apex of the berry carries a central scar, with four tiny symmetrical scars around it.

The berries are soft and mucilagenous, the extremely sticky nature of the pulp causing the seeds to adhere to the beaks of birds as they eat them. Mistle-thrushes are very fond of them, stripping the plants long before Christmas time. After feeding, they can be seen vigorously wiping their beaks on the tree bark to remove the sticky pulp; in this manner, seeds are spread to other trees. Each fruit contains a single seed, which is white, heart-shaped and flat, with a blunt apex. The berries are ripe from November until February.

*Leaf and Flower*

Mistletoe is a bushy perennial, semi-parasitic on the trunks of trees, with specialized organs called 'haustoria' which penetrate the tissues of the host plant and extract part of the nutrients of the mistletoe. Stems grow out of the bark of the host up to 75cm long, with repeated dichotomous branching. The stems are green with sparse fine short hairs on the immature twigs. The yellow-green, bluntly lanceolate leaves are sessile, in opposite pairs and leathery in texture. Flowers are produced from late February to early May, male and female on different plants, in small clusters. The male flowers 4mm in diameter have no calyx but four sepaloid triangular petals on which grow four sessile anthers. The female flowers, 2mm across, lie in a cup-shaped bract, having a small indistinctly four-toothed calyx and four tiny petals. Recent research has shown that the flowers are frequently pollinated by the Green-bottle (*Dasyphora cyanella*) and less often by smaller flies. These feed on nectar and pollen, the pollen grains being large and unsuitable for wind dispersal.

*Ecology and Distribution*

Mistletoe is locally common as a parasite of various trees in England and Wales, north to Yorkshire, most frequently in the west Midlands and south-east England, and it is a characteristic plant in parks and orchards of the Severn Valley in Herefordshire, Gloucestershire and Worcestershire. It is absent from Scotland and Ireland, and it would seem that altitude is a limiting factor in its distribution since it does not occur above 300m.

McClintock describes three races of mistletoe, the commonest being parasitic on broad-leaved trees, chiefly apple, poplar and plum, rarely on oak and very rarely on pines, larches and other firs. Deakin (1871) also recorded parasitism on hawthorn, common whitebeam and hazel, while Cove (1956) found that out of 831 infestations 345 occurred on apple, 222 on hawthorns, 133 on elm and 76 on lime. Within twenty miles of Hereford mistletoe has been recorded on thirty host species, including apple, hawthorn, poplar, lime, false-acacia and field maple. Severe parasitism can lead to the death of the host tree.

Mistletoe occurs in most of Europe except the northern and eastern borders, and anyone who has driven down the long, straight roads of north France cannot have failed to see the round bunches growing in the poplar trees which line the roads.

*History and Uses*

The history of man's association with mistletoe is ancient, and there is evidence of the involvement of mistletoe in druidic and prehistoric rites, mistletoe found on oak being especially prized. It features in Norse mythology in the legend recounting the murder of Balder, the god of Light. In Saemund's

**Mistletoe**

*Edda* the story tells how, when the death of Balder is foretold, Odin his father and Frigga his mother take steps to obtain a solemn vow from all things animate and inanimate that they will not harm their son. Only the mistletoe is excepted, since it seems too puny. Loki, the evil god of Fire, by magic arts causes the mistletoe to grow long and hard, and from it fashions a shaft which he gives to Balder's blind brother Hodur, the god of Darkness. The gods, in play, are showering missiles on Balder, who stands smiling and unharmed, until the mistletoe shaft, thrown by his unwitting brother, strikes him dead.

Mistletoe berries have been widely used in herbal medicine for treating epilepsy and chorea, asthma, infertility and cancers, although in the latter case there is no basis to expect it to be of any value. It is said to be diuretic and spasmolytic, and has been used in treating high blood pressure and arterio-sclerosis. Culpeper noted: 'The bird-lime which is made of the berries of Misseltoe is a powerful attractive, and is good to ripen hard tumours and swellings'. Gent (1681) voiced the same opinion in the treatment of horses, and added 'Misseltoe

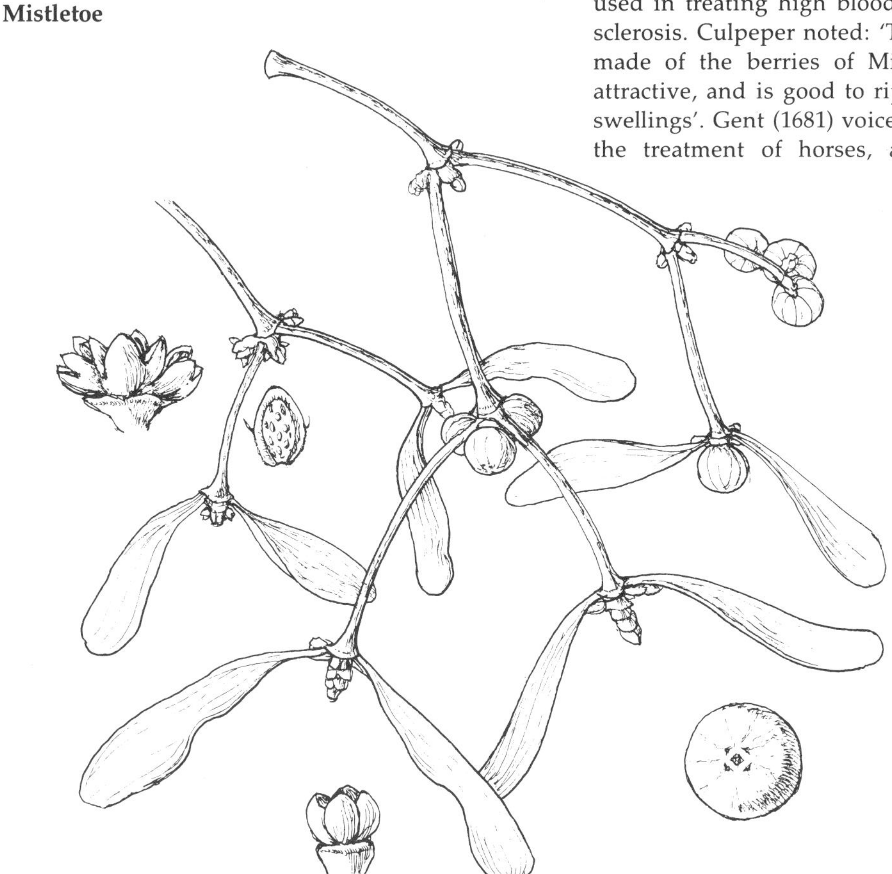

bruised, and the Juice put in the Eares, healeth the Impostumes in them in a few days. The Misseltoe of the Oak being given inwardly, Cures the Falling Sickness, or hung about his Neck'.

*Toxicology*
The leaves, stems and seeds of mistletoe contain approximately eleven viscotoxins (a mixture of closely related polypeptides) although these are absent from the flesh of the berries. Among the polypeptide albumins are viscotoxin A, tyramine, β-phenylethylamine and arginine. Viscin and viscunin (lectins similar to ricin), are present in small quantity, with acetylcholine and proprionylcholine. Mistletoe acts as a gastric irritant and is also cardiotoxic and sympathomimetic, although the vasodilator action, formerly used to reduce high blood pressure, is delayed. The maximum effect occurs three or four days after ingestion, which should be borne in mind when treating acute poisoning.
**Symptoms** The consumption of three or four berries by a child, or ten by an adult, will produce nausea, vomiting and diarrhoea; there is pallor and the pupils are dilated. Severe poisoning causes gastroenteritis with colic and slowing of the heart rate. Muscle weakness is experienced, also hallucinations and marked diuresis.

Mistletoe is still included in herbal remedies, and there is a recent report (Harvey and Colin-Jones) of a preparation containing mistletoe causing hepatitis in a woman, who showed symptoms of nausea, malaise and a dull ache in the upper abdomen. There are recorded cases of poisoning in dogs and horses: a griffon which ate mistletoe berries suffered paralysis with opisthotonus, a rapid pulse and respiratory rate, and died fifty hours later. Another dog which ate mistletoe showed abdominal discomfort, dilated pupils and a slow respiratory rate, but recovered spontaneously. Symptoms in the horse included colic, incoordination and difficulty in breathing.
**Treatment** It would be wise immediately to seek medical help for any child or animal who has consumed more than three or four berries. With children symptomatic supportive care may be needed, with administration of fluids and control of the bradycardia with atropine. In cases of severe poisoning ECG monitoring of the heart is wise and in the USA disodium-edetate has been used to good effect. Although mistletoe is widely available in the home at Christmas time, few berries are taken because of the bitter taste, and reported cases of poisoning are rare: most of the anecdotal cases dating back before 1900.

# · *BERRY CATCHFLY* ·

**Cucubalus baccifer** L.

? POISONOUS

*Fruit*
Berry catchfly is unique among European Caryophyllaceae (Pink Family) in having a fruit which is a fleshy berry. These berries are spherical, black and shining, 6-8mm in diameter, drooping when ripe on pedicels 10-15mm long. The skin is tough but rather brittle, like thin plastic, enclosing a layer of soft green pulp within which lie forty to fifty seeds. The green sepal remnants are usually still present at the base of the fruit, which in a sunny locality will ripen from July onwards. In more shaded places ripe fruits may be found as late as November, but in England it appears to flourish better in a warm summer. Finally the ripe berry shrivels and bursts, dispersing black, kidney-shaped shiny seeds 2mm long, each with a peg-like knob at the hilus. The berries are attractive to birds, which may be the principal agents in spreading the species; blackbirds, whitethroats and great tits have been recorded eating the fruits freely.

*Leaf and Flower*
Berry catchfly is a straggling, hairy perennial, which will grow to more than 1m high, scrambling through supporting vegetation. The foliage has a rank smell very similar to that of solanaceous plants such as bittersweet (*Solanum dulcamara*). The thin, round stems are rather brittle; in contrast the pale

green leaves are pointed oval and soft in texture. The stem, petiole, leaf edges, and larger leaf veins bear fine, appressed hairs. The lower leaves are short stalked, and the upper leaves sessile. The whole plant resembles a broad-leaved, scrambling bladder campion (*Silene vulgaris*). Berry catchfly flowers from June to September. The sepals are joined into a short, bell-shaped calyx tube, covered with stiff hairs and with the ends of the lobes reflexed. There are five creamy green petals which are deeply cleft, with scales above their spreading narrow lobes. There are three styles. As the fruit ripens the petals shrivel and the calyx shrinks back to expose the berry.

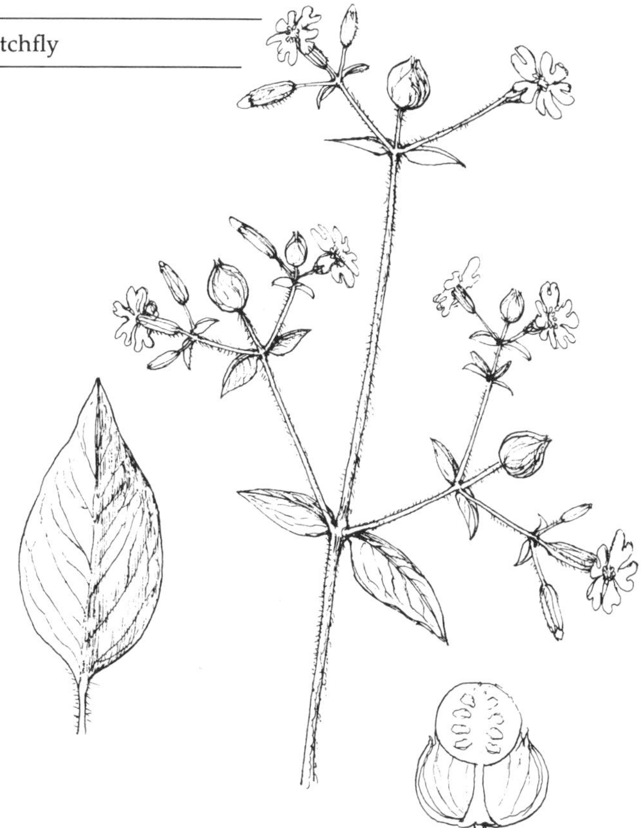

*Ecology and Distribution*

Berry catchfly grows in Britain in roadside hedges, in the semi-shade of moist woodland, and along river banks where the ground is moderately damp and rich in humus. It is a very rare plant of possible native status, occurring with regularity only in a few stations in Norfolk (in the parishes of Merton and Hockham) where one site is in dense fen vegetation. The classic site was on the banks of a ditch in the Isle of Dogs in Middlesex, where it

**Berry catchfly**

flourished from 1837 to 1853. J.E. Lousley considered it undoubtedly to be native in Middlesex and Norfolk. There are records for Woody Bay in west Devon in 1921, and in a hedge at La Trinité in Jersey in 1930, where it may have been a garden escape. More recently a single plant was found near Royal Portbury Dock, Somerset in 1978; and another discovered in a lane hedge near Langport, north Somerset in 1981.

In Europe, berry catchfly is most common in south and central areas, north to the Netherlands and western Russia, where it is usually recorded from damp hedges near river banks and in moist woodland.

*History and Uses*

Berry catchfly has a long history in Britain, being first recorded as a British plant in 1570. It was cultivated widely by herbalists in the early 17th century. Culpeper called it Berry-bearing Solanum (*Solanum Bacciferum*), writing that 'It is not poisonous, but is an excellent counter-poison, and an alexipharmic, good in malignant and pestilential fevers'. Gerard claimed the virtue of berry catchfly as an emollient poultice.

*Toxicology*

Berry catchfly belongs to the *Caryophyllaceae*, a family which contains species known to be toxic. The toxic principles are usually saponins with an irritant action on the stomach and intestine, and possibly a depressant action on the heart and central nervous system.

**Symptoms** Ingestion of large quantities of berries might lead to gastroenteritis and slowing of the heart rate. There are no references to any cases of poisoning in man or animals in Britain, and in view of the rarity of the wild plant none may be expected.

**Treatment** Vomiting should be induced if it has not already occurred, and 5-10g of activated charcoal should be administered by mouth.

# · *BANEBERRY* ·

## *Actaea spicata* L.

POISONOUS

*Fruit*

The fruit of baneberry is formed from a single carpel, ripening from green to a glossy black berry 10-13mm long. In shape it resembles a rugby football, a likeness enhanced by the shiny leathery skin of the berry and the deep grooves down each side. The fruiting spike elongates as the berries ripen; Salisbury records a plant bearing as many as 240 berries. However, the attachment of the berry to the pedicel is extremely brittle, and where it grows on exposed limestone pavement, most of the berries fall off before becoming fully ripe. Each berry contains an average of eight seeds, and they are ripe in September.

*Leaf and Flower*

Baneberry is perennial and almost hairless, with glossy green leaves on long slender stalks, each leaf twice trifoliate, with acutely toothed leaflets. By the time the fruits are ripe the leaves are usually scorched brown and shrivelled at the edges. The whole plant is 30-65cm high, with creamy white flowers in a dense, short raceme. Each flower is made up of three to five petaloid perianth segments and four to ten small honey leaves which are shorter than the stamens. The numerous white stamens protrude from the flowers, so that the whole spike has a frothy appearance. Baneberry flowers in May and June.

*Ecology and distribution*

Baneberry is a rare and local plant of limestone pavement and calcareous ash woods, growing on the carboniferous and magnesian limestones of the north of England, at elevations up to 500m. It is found in about sixty stations in Yorkshire, although in two of these it is present as a single plant. There are two stations in West Lancashire, at Ease Gill and Leck Fell, and a further station in Westmorland. The Leck Fell site is on an inaccessible ledge, and

**Baneberry**

was confirmed in 1965 with the use of binoculars! In many localities where it grows deep in the grikes (fissures) in the limestone pavement, baneberry is easily overlooked; even when found, it is awkward to photograph, as the berries are deep in a cleft below ground level. However, at Folkton Brow in East Yorkshire it dominates the ground flora of the ashwood over a considerable distance. It is widespread on the scars and pavements of Upper Wharfedale and Littondale, a fascinating area which has a desolate beauty all of its own.

Baneberry is widely distributed through most of Europe, but in the south it is restricted to calcareous mountains.

*Toxicology*

All parts of the plant are poisonous (hence its name). The toxic principle, protoanemonin, is irritant and causes a burning sensation of the lips and tongue, followed by violent gastroenteritis.

**Symptoms** In humans poisoning produces headache, dizziness and hallucinations, with a rise in pulse rate and possible shock. No cases of poisoning in animals have been recorded.

**Treatment** The patient must be made to vomit, and gastric lavage may be necessary, with supportive care. If severe symptoms are present, 2 mg atropine should be injected subcutaneously, and repeated as necessary. Poisoning with baneberry is rare, since the plant is highly unpalatable. NPIS record four human cases in recent years; the symptoms were seldom serious.

# BARBERRY

**Berberis vulgaris L.**

POISONOUS

*Fruit*

The fruits of barberry are brilliant orange-red. A heavily laden bush is a most attractive sight, and appears to be dripping with fruit from top to bottom. The berries are in drooping racemes of twenty to forty on reddish brown pedicels 6mm long, each pedicel slightly swollen below its attachment to the fruit. The berries are yellow and slightly curved when unripe, resembling tiny bananas, with the black remnants of the flower just off-centre at the top. They ripen from late July to September, becoming fatter, 7-12mm long and 6mm in width. The wall of the shiny berry has the texture of a soft rose hip, and within the single cell are two or three long oval seeds, 7 × 3mm, shiny pale brown, slightly pitted, and flattened on one side. The berries have a sharp acid flavour which is not unpleasant. In dry, cold summers they may not ripen, but remain thin and yellow well into August, before falling off.

*Leaf and Flower*

Barberry is a deciduous shrub up to 4m high. It has glossy oval leaves with toothed edges, and grooved twigs armed with three-pointed spines. If the bark is peeled off it reveals bright yellow wood beneath. The yellow, sweetly scented flowers appear in May and June and are borne in drooping racemes. Each flower consists of three outer perianth segments, small and sepal-like, and six bright yellow, petal-like inner perianth segments. The honey leaves are also petal-like, and the whole flower is 6-8mm in diameter. The flowers secrete abundant nectar which is particularly attractive to bees. The anther filaments are elastic, and when touched by an insect they contract suddenly, closing round it and dusting it with powdery pollen. This process can be repeated.

*Ecology and Distribution*

Barberry is native to Europe, growing widely ex-

cept in the extreme north and in the Mediterranean. It is possibly native in a few places in England, but elsewhere has been introduced. The limits of its natural range are hard to determine, for it was widely planted both for its ornamental flowers and for its edible fruits. It is most commonly found in Britain as a hedgerow plant. Several other species of *Berberis* have escaped from cultivation and established themselves locally.

*History and Uses*

In the Middle Ages barberry was much planted for its berries, which were used for making an excellent dessert jelly, and for its bark, which was used as a remedy for jaundice. The juice was used in the treatment of disorders of the liver and gall-bladder, and Culpeper remarked that 'the berries are as good as the bark and more pleasing; they get a man a good stomach to his victuals . . .' Gent (1681) wrote that 'the berries are cold and dry in the second degree, and as Gallen affirmeth, are of thin parts and have a certain cutting quality; they are good to stop Lasks and bloody Fluxes.' The bark and wood yield a bright yellow dye, and were used in Poland for tanning leather. The concentrated juice, rich in vitamin C, can also be preserved and used in a similar manner to lemon juice. When using the berries for jelly-making, care should be taken not to crush the seeds.

Barberry acts as the secondary host of black stem rust (*Puccinia graminis*), of wheat, and overwinters on it. For this reason it has been extensively destroyed, and in many areas it is a rare and decreasing species.

Nine species of birds have been recorded eating the berries.

**Barberry**

*Toxicology*

The ripe fruits are edible, but the unripe berries are poisonous. The toxic alkaloid berberine is present with small quantities of oxyacanthine and chelidonic acid, mainly in the bark of the roots and the unripe fruits. Berberine has been used orally as a bitter and in the treatment of cholera in doses up to 150 mg daily. Injections of berberine sulphate have also been used in India in the treatment of cutaneous leishmaniasis.

**Symptoms** Berberine acts as a gastric and mucosal irritant and will cause vomiting and diarrhoea if very large quantities are ingested. It may also depress respiration through action on the central nervous system.

**Treatment** Vomiting should be induced and supportive care given if indicated. NPIS have recorded fewer than ten cases, which all exhibited minor symptoms with no untoward sequelae. No authenticated cases of poisoning in animals have been reported in Britain.

# · OREGON-GRAPE ·

***Mahonia aquifolium*** **(Pursh) Nuttall**

POISONOUS

*Fruit*

The fruits of oregon-grape are indeed grape-like, small and round, 8-9 mm in diameter, blue-black, with a fine bloom on the skin of the berry. The fruits are borne in loose clusters of up to fifteen on pedicels 6-7 mm long, each springing from the axil of a small persistent ear-shaped green bract. They are ripe in August and September, and have a pleasant sweet flavour when eaten raw. The flesh is reddish purple, with a strongly purple staining juice, and most berries contain three or four shining yellow, smoothly triquetrous seeds, 4 × 1.5 mm.

*Leaf and Flower*

Oregon-grape is an evergreen shrub up to 1.5 m high, spreading by stolons and with spineless stems, unlike barberry. The leaves are pinnate, with

**Oregon-grape**

leaflets in opposite pairs and a terminal leaflet; as the scientific name suggests, they resemble the leaves of holly, having large teeth armed with spines. The leaves often take on a bronze tinge in winter, but there is in any case a considerable variation in the degree of lustre of the leaf surface, also in the breadth of the leaflets and the coarseness of the toothing. Many garden plants are the hybrid *M.aquifolium* x *repens*; the Californian *M.pinnata* (Lag.) Fedde, with dull, finely serrate leaves is also in cultivation. The sweetly scented yellow flowers are carried in dense upright spikes, each flower 8mm across, formed of three outer and six inner perianth segments in a densely packed little cup. The flowering period is normally from March to May, but some shrubs will flower in January.

*Ecology and Distribution*

Oregon-grape was introduced to Britain in 1823 from the north west of North America, and is most frequently grown in shrubberies, shelter belts and game coverts. It has become widely naturalized in central and east England, and occurs as far north as East Perth, but not in Ireland. This species or one of the many other *Mahonia* species or hybrids may be encountered anywhere in Britain, as they are commonly used in landscape planting. They are similarly cultivated and locally naturalized in west and central Europe.

*History and Uses*

The original reason for growing oregon-grape was as a shelter and feed for pheasants and other game birds, though wild birds will also enjoy the ripe fruits. The berries have a sweet tangy taste and make an excellent jelly, though gathering enough to be worthwhile is a prickly task.

*Toxicology*

The ripe fruits are edible, but, as in the case of the barberry, the bark and roots contain a very small amount (0.06%) of an alkaloid which affects the central nervous system, and can cause respiratory depression. No case of poisoning of man or animals has been recorded in Britain.

# · RED CURRANT ·

**Ribes rubrum** L.

Red currant is widely cultivated for its fruit and may not be truly native to Britain.

*Fruit*

The berries of red currant are round, 6-10mm in diameter, crowned with the shrivelled flower remnants, and borne in drooping, stalked spikes of ten to twenty, although in the wild, birds strip the bushes as soon as the fruits ripen. Ripe red currants are juicy, with a translucent skin through which the seeds are visible. The berry is well marked with ten green ribs which radiate from top to bottom like the lines of longitude on a globe. The seeds, 3.0 × 3.5mm, are pointed oval and yellow, with a honeycombed surface. They have a sweet, rather acid, flavour and are ripe in July. White berried forms are not uncommon and have a sweeter taste; when ripe the seeds are clearly visible through the skin of the berry. White berries may also have a slightly flattened, quoit-like shape, and the berry pulp is often tinged red.

*Leaf and Flower*

Red currant is a deciduous shrub, 1-2m high, with erect but rather straggly brown stems which may be hairy. The long-stalked palmate leaves have three to five bluntly toothed lobes, and a deep, narrow sinus at the base. They are hairless, except on the lower surface of the midrib, and scentless, which immediately distinguishes them from the strongly aromatic leaves of black currant (*Ribes nigrum*), even where flowers or fruits are absent.

They flower from late April to mid May. The small flowers, 5-6mm in diameter, are borne in drooping racemes of ten to twenty from the leaf axils. Each flower has four or five blunt sepals,

green or tinged with red, and four or five very small cuneiform green petals. The receptacle is saucer-shaped with a raised, five-angled rim. There are five stamens, and the flowers are hermaphrodite. *R.spicatum* Robson is closely related to *R.rubrum*; it has larger leaves with a less defined sinus at the base, and the leaves are hairy beneath. The flowers are similar, except that there is no raised edge to the receptacle.

*Ecology and Distribution*
Red currant is widespread throughout England, Wales and Scotland, especially in south and east England. It is absent from Orkney, Shetland and the Hebrides, and very scarce and local in Ireland, where it is almost certainly an introduction. It is a plant of river banks, wet woods, fen carr, and wet alder flushes, where it may be accompanied by black currant. It grows at altitudes up to 450m.

*R.spicatum* is a very local and decreasing plant of limestone woods in the north of England, in Yorkshire, Durham, Lancashire and Cumbria, and in Scotland north to the Moray Firth. It grows elsewhere in north and east Europe, and is rarely naturalized outside its native area.

*History and Uses*
Red currants are edible, but because of their acidity they are best used mixed with other early summer soft fruits, especially raspberries, to which they impart a refreshing tang. They make excellent jelly. The wild berries are seldom found in any quantity, as they are eaten with avidity by birds and small mammals as fast as they ripen. The white-berried form seems to be less attractive to birds, and large sprays of ripe fruits can often be found untouched by predators. Red currants have no special medicinal use.

**Red currant**

# · BLACK CURRANT ·

**Ribes nigrum** L.

*Fruit*
The berries of black currant are shiny black, globular, 8-15mm in diameter. The apex of the berry is crowned with the shrivelled floral remnants. Most wild black currants are nearer 8mm in size, much smaller than the garden cultivars. The internal structure and ribbing of the berry is not clearly visible as it is in red currant. The berries are borne in drooping sprays of three to ten, the lowest fruits in the spray having the longest pedicels. They have a strong aromatic smell, which has some resemblance to that of tom cats, *but* the fruit has a delicious, sweet and tangy taste. The juice is strongly purple-staining. Each berry contains numerous ovoid seeds, yellow in colour and marked with vertical ribbing. The fruits are ripe in July.

*Leaf and Flower*
Black currant is a stout, erect, deciduous shrub, growing to 2m in height. The palmate leaves are coarsely serrated and have a deep, wide sinus at the base. On the veins and midrib of the undersurface of the leaf there are fine hairs and sticky glands which give the foliage a strong scent even more cat-like than the fruit. Black currant flowers from late April to mid-May. The small flowers, 8mm across, are borne in drooping racemes of three to ten. The bracts are short, hairy and less than half the length of the pedicels. The sepals are erect and recurved at the tip, making a bell-shaped calyx which bears glandular hairs. The whitish petals are ovate, only half the length of the sepals, and the same length as the stamens.

*Ecology and Distribution*
Black currant is frequent in damp woods, by streams, and in fen carr throughout Britain, especially in the south and east. It is far less common in Wales and north Scotland. In Ireland, where it is introduced, it is scarce. It grows in most of Europe except the Mediterranean, and is certainly native to central and east Europe. Elsewhere it may be na-

tive, but it has been cultivated for so long, with inevitable escapes and naturalization, that the picture is far from clear. Birds and small mammals are quick to eat the berries as soon as they ripen, as any gardener will know, so that it is not easy to find wild fruit in any quantity.

*History and Uses*
Black currants have been cultivated for a very long time for the excellence of their fruit as dessert and for making into jams, jellies and syrups. Many of the old herbalists recommended jellies and lozenges containing black currant juice for the relief of sore throats, a use which still continues. An 18th-Century herbal refers to black currants as 'Squinancy Berries', because of their use in treating quinsy, or 'squinancy'. The berries contain 6-8% invert sugar, and have an ascorbic acid (vitamin C) content of 100-300 mg/100 g of fruit. Black currant is used as a rich source of vitamin C, as a flavouring agent, also as a diuretic and in the treatment of gastritis and diarrhoea. Deakin (1871) observed that: 'In Siberia, the leaves dried and mixed with Souchong tea give it a flavour of green tea, and when infused in spirit they give it the colour of common brandy'.

**Black currant**

# · GOOSEBERRY ·

***Ribes uva-crispa* L.**

The origin of the name gooseberry is not clear; it is not necessarily a corruption of Dutch, German or French names: the application of animal names to plants and fruits is sometimes inexplicable. In 16th-Century writings it is referred to as Groseberry and Gozeberry, with many other variations.

*Fruit*
The fruits of the wild gooseberry are much smaller than those of the cultivated form, being 10-20 mm long and 10-12 mm in diameter. They are usually borne singly on a hairy pedicel 5 mm long from the leaf axils. By mid-June fruits are well formed, but they do not ripen until July, when they are still green, sometimes tinged with yellow or red. The scientific name *uva-crispa* derives from *uva*, the Italian and Spanish for grape; and *crispa*, meaning hard. The berry may be smooth, but it is normally stiffly hairy, with the shrivelled floral remnants at the apex. The radiating green ribs, usually ten, run from top of bottom of the berry, as in the berries of red currant. The flesh of the gooseberry is translucent green, the berry wall 1.5 mm thick, being divided vertically into two compartments. In each compartment there are four rows of seeds, numbering twenty to thirty in all. The seeds are brown and

Gooseberry

wrinkled, bluntly angled and about 3 mm long, each enclosed in a sheath of transparent jelly. Wild gooseberries are very sour-tasting, with little flavour.

*Leaf and Flower*

The gooseberry is a small, spiny, deciduous shrub up to 1.5 m high, with many hairless stems armed with groups of one to three sharp spines at the nodes. In winter the twigs have a white, papery appearance. The leaves are small, cordate, with three to five bluntly toothed lobes, smooth above and slightly hairy on the underside. The flowers are carried in short-stalked racemes of one to three, appearing from as early as mid-March, until May. The bracts are very small, and the drooping, campanulate flowers, 10 mm across, have oblong green sepals which are often tinged red around their edges. The petals are paler, tiny and erect, and the stamens are twice the length of the petals.

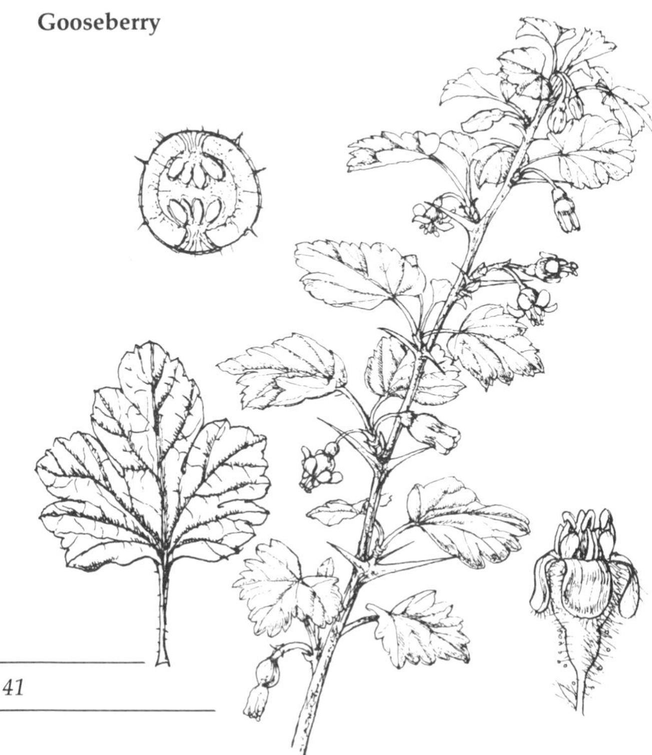

*Ecology and Distribution*
The gooseberry is widely cultivated but grows wild in woods and hedgerows and by streams throughout England, Wales and Scotland. Although widespread, it is not frequent, being absent from Orkney, Shetland and the Outer Hebrides, and very local in Ireland, where it has been introduced. It is native to southern, central and western Europe, including mainland Britain, but is also extensively naturalized. Flowering is sparse in woodland plants, and fruiting is even less common, with plants flourishing best where there is some sun and not too much competition from other flora. An epiphytic gooseberry bush has been seen growing from the fork of an ash tree at Jennie's Foss near Malham, and K-H. Loske reports gooseberry as an epiphyte on both poplar and willow in Germany. It is a shrub which is very easily overlooked. Birds and small mammals will eat the berries, which also seem attractive to slugs and snails.

*History and Uses*
Gooseberries are best cooked before eating, making excellent pies, jams and preserves. The addition of elder flowers to gooseberry jam imparts a particularly delicious flavour. Culpeper wrote: 'The ripe Gooseberries are excellent to allay the violent heat of both the Stomach and Liver'.

# · *MOUNTAIN CURRANT* ·

***Ribes alpinum*** **L.**

*Fruit*
The berries of mountain currant are globose, translucent shiny red and carried in erect spikes, which distinguishes them from the drooping spikes of red currant. The berries are small, 6-10mm across, crowned with the shrivelled remnants of the flowers, and very sweet tasting, with no trace of acidity. They ripen in July and August, and most berries contain two or three relatively large yellow ovoid seeds, 3 × 1.5mm, which are vertically ribbed and irregular in outline.

*Leaf and Flower*
Mountain currant is a deciduous shrub up to 3m tall. It grows taller, thicker and more branched than red currant, with darker, smaller leaves which tend to have three narrowly toothed lobes rather than five. The leaves are hairless on the underside, bear a few stiff white hairs on the upper surface, and are scentless. Flowers appear from late April to the end of May, five to twelve in an erect raceme with prominent green bracts; 6mm in diameter, they have five spreading green or red tinged sepals, with

**Mountain currant**

five tiny cuneiform petals set between them. The receptacle is round and saucer-shaped.

Mountain currant is dioecious, with male and female flowers on separate bushes, but each has rudimentary sexual structures of the opposite sex. It does not bear fruit very well; arguably this may be because of the relative scarcity of male plants. Another reason for the lack of fruits may be the absence of the correct insect pollinator.

*Ecology and Distribution*

Mountain currant is a very local plant of river banks, cliffs and rocky woods on the carboniferous limestone of North Wales and the north of England. The scrubby woods where it grows are composed chiefly of ash, with wych elm and common whitebeam. It is native to Denbigh in North Wales, the Derbyshire Dales – especially Dovedale and Millers Dale – and the northern and western dales of Yorkshire; it is frequent in the woods around Rievaulx. Elsewhere it has been widely introduced, mainly in northern and western Britain, in such places as High Force in Teesdale, Ben Vrackie near Pitlochry. It is an established alien in the Lothians in south east Scotland. In one particular Derbyshire dale the valley floor is carpeted with alternate-leaved golden-saxifrage (*Chrysosplenium alternifolium*), while the cliffside is dripping with mountain currant.

In Europe it grows mainly in the northern and central areas, extending southwards in the mountains to northern Spain, central Italy and Bulgaria.

*Uses*

Although the berries of mountain currant are delicious to eat they should not be gathered for that purpose in Britain, since it is an uncommon plant.

# · CLOUDBERRY ·

***Rubus chamaemorus*** **L.**

*Fruit*

Cloudberry is a collection of large drupelets numbering four to twenty, the whole fruit having a diameter of 15-17 mm, carried singly, upright on long pedicels. When unripe the berry is hard and red, with the calyx folded around it, but by the time it has ripened in late July and August the large pale calyx lobes fold downwards and the fruit, juicy but firm in texture and with a distinctive flavour, turns pale pinky-orange. Each drupelet contains one rough, hard stone 4mm long, brown and slightly triquetrous.

*Leaf and Flower*

Cloudberry is a low, downy, creeping perennial 5-20 cm tall, forming open patches of growth. The leaves are large and rounded, up to 8 cm across, palmately lobed with blunt serrations, carried on

long slender stalks. When the leaf buds first open in spring they are shaped like little fans, and at that stage bear glandular hairs. When fully expanded they resemble the leaves of mulberry. The flowers are solitary on erect pedicels, white, 15-24mm across, with four or five petals much larger than the sepals, resembling large blackberry flowers. They appear in June and early July before the leaves are fully developed, male and female flowers on different plants, the male flowers having numerous stamens with dark anthers. When the petals are shed the calyx of the female flower folds around the developing fruit.

*Ecology and Distribution*

Cloudberry is locally common on damp mountains and heather moors in the north central and north west Highlands, favouring deep peat bogs and heather-covered peaty corries at altitudes varying from 180m in Sutherland to 1140m on Beinn à Bhuird. It grows with *Eriophorum vaginatum*, *Vaccinium uliginosum* and *Calluna*, on level or gently sloping hummocks. It is absent from Shetland. In North Wales it grows in the Berwyn mountains on the borders of Montgomery, Merioneth and Denbigh, where it is locally abundant in blanket bogs above 800m. In northern England it is common in the blanket peat of the north east Pennines, growing best in hummocky, well-drained areas. In the southern Pennines it is a local plant, growing at higher levels on the cotton-sedge mosses. It also occurs, rarely, in the Lake District. It is very rare in Ireland, being much decreased in its only site in the northern mountains of Tyrone.

The male flowers secrete abundant nectar, and insect pollination certainly occurs despite the dearth of insects on the cotton-sedge moors where cloudberry grows. W.H. Pearsall recorded a peculiar *Syrphid* fly, a melanic form of *Melanostoma mellinum*, as being characteristically associated with cloudberry in the Pennines and west Highlands, and also two Empids, *E.lucida* and *E.snowdoniana*. Despite this, reproduction is mainly vegetative, the plant spreading by a creeping, much branched rhizome up to 1.7m long. J. Raven remarks that in the south Pennines the apparent sterility (lack of fruit) is a reflection of the greater frequency of male clones, and that the male and female plants may be reacting differently to climatic changes, the female flowers being more susceptible to adverse conditions. Plants may not fruit more than once every seven to twelve years, and natural seedlings are rare, taking at least five years to reach flowering maturity. The flowers, which have a high nutritional value and stimulate milk flow are cropped by deer and hill sheep, and this grazing leads to poor flowering and fruit set. Moor-burning in May often leads to an increased fruit crop in the following autumn, but the berries should be gathered before they are stripped by red grouse, which love them.

Cloudberry grows in northern Europe, especially in Norway and Sweden, and reaches as far south as Czechoslovakia and southern Germany, although it is a true Arctic species.

*History and Uses*

De l'Ecluse in his *Stirpium Pannonicarum Historia* wrote that he was sent cloudberry in the late 1500s

**Cloudberry**

by Thomas Penny, who found it 'growing in great plenty, among heather on Mount Ingleborrow, the highest in all England'. M. Walters in *Mountain Flowers* recounts an interesting historical reference from the village of Llanrhaiadr-ym-Mochnant in the Berwyn mountains in Wales. There the berries are called Mwyar Dogfan, the tradition being that a quart of cloudberries were the wages of Dogfan Sant for his care of the parish. The custom persisted that if anyone could present a quart of cloudberries to the parson on the morning of the saint's day his church taxes for the year would be waived. Cloudberries are delicious to eat, but it is rarely possible to gather them in any quantity in Britain. In Norway and Sweden they fruit readily, and are extensively harvested for use in jam, preserves and wine; cloudberry yoghurt is commonly available in the supermarkets there, and is well worth buying. The Swedish name for cloudberry is Hjötron. A liqueur made from them can be purchased occasionally in Scandinavian shops in Britain; it should be served well chilled.

In the north Yorkshire Dales the name for cloudberry is 'Naughtberry', hence Naughtberry Hill east of Buckden Pike. As the ripe fruit is soon eaten by the red grouse, people too often gather the berries when they are red and underripe and truly taste of naught!

# · *STONE BRAMBLE* ·

***Rubus saxatilis*** **L.**

*Fruit*

Stone bramble is a shy flowerer, and it fruits even less readily, the berry when ripe resembling a depauperate red blackberry. The fruit is a collection of drupelets numbering only two to six, each 5-6mm across, which are translucent and scarlet when they are ripe from mid-July to September. Each drupelet contains a single large oval stone, 4 × 2mm with a pitted surface. The fruits are edible but rather insipid, being composed mostly of stones, with very little juice.

*Leaf and Flower*

Stone bramble resembles a slender, trailing blackberry with thin downy stems armed with delicate straight prickles. The stems, which die back each winter, can be more than 1m long, and creep about in the cracks of rocks. The plant also has long above-ground stolons which root at the tips. The leaves are trifoliate, with long, slender stalks. The leaflets are unevenly toothed, the upper surface being smooth and the paler under surface felted with fine down. The flowers are small, 6-10mm across, dirty white, with narrow, erect petals surrounded by conspicuous green sepals which reflex

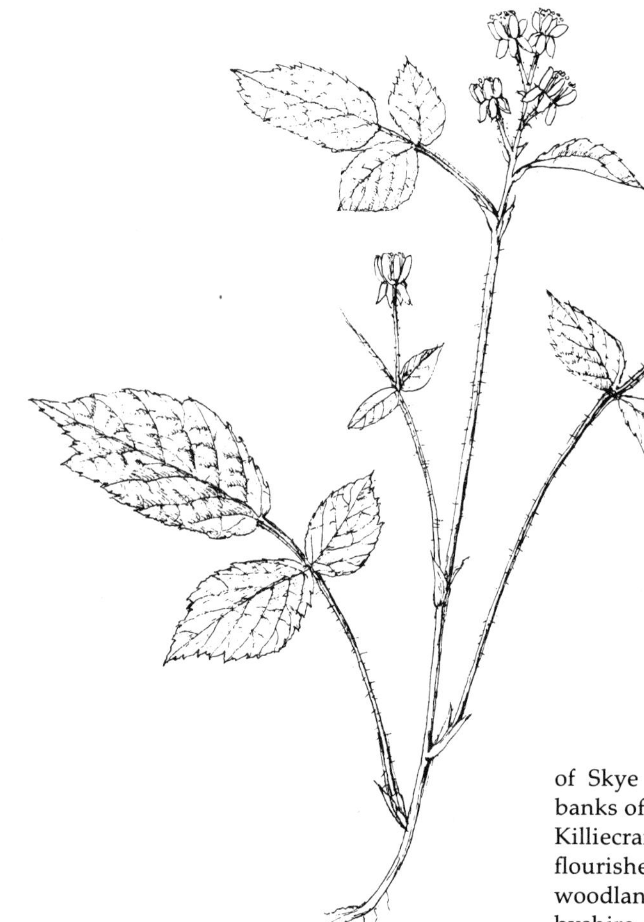

**Opposite: Stone bramble**

in fruit. The flowers are in loose clusters on long thin pedicles, and are pollinated by bees and flies, or self-pollinated. This bramble likes moist, shady places among the rocks, and it is quite a problem to find plants bearing flowers, even more so to find ripe fruit.

*Ecology and Distribution*

Stone bramble is widespread but local on rocks and ledges in the north and west of Britain, especially on hills in limestone districts, although recently it has decreased in Ireland. It occurs widely on calcareous and base-rich rocks in the north and north west Highland, in Orkney, Shetland, the Outer Hebrides and on the strongly basic plutonic rocks of Skye and Rhum. It also grows on the gravel banks of Highland rivers such as the River Garry at Killiecrankie and the River Awe at Taynuilt. It flourishes in the damp carboniferous limestone woodlands of Yorkshire, Lancashire and Derbyshire, and in Wales in two main areas, the mountains of Snowdonia in the north and the Brecon Beacons in the south. It has a single southern site in north Devon, and in Ireland is widespread on the limestones of Co Clare, Galway and Limerick. It occurs in most of Europe, but is rare in the south west.

*History and Uses*

The fruits of stone bramble are edible, but they fall readily when ripe, and the whole plant, delicate and creeping among rocks and ledges, is easily overlooked. There is an interesting reference to *Rubus saxatilis* in Charles Dickens' *The Miner's Daughters,* in which he describes children walking on the hills near Wardlow Dale in Derbyshire, where they pick 'clusters of the mountain-bramble, resembling mulberries and known only to the inhabitants of the hills'.

# · RASPBERRY ·

***Rubus idaeus*** **L.**

*Fruit*

The sweet, red fruit of the raspberry is a drupe 10-20mm long, 10-15mm wide, composed of a mass of drupelets. Fruit is carried in drooping clusters near the tips of the year-old stems, each fruit on a pedicel 10-20mm long which bears small prickles. Below the fruit the prominent pale green calyx lobes are reflexed, and the ripe fruit separates easily from the long conical receptacle. When the fruit is removed, the remnants of the stamens are left as a hairy collar around the base of the pale receptacle. The fruit is fleshy and sweet, with red juice. Each drupelet is 3.5mm long × 2.5mm wide, finely downy and crowned with the stylar remnant, which projects like a crooked whisker from the apex. Inside the drupelet is a rough brown seed 2.0 × 1.5mm. White – or golden-berried – forms of raspberry are not uncommon, and have a distinctive, winey flavour. Wild raspberries are ripe in late June, July and August.

*Leaf and Flower*

Raspberry is a little-branched perennial 1-3m high, suckering freely by adventitious buds from the roots, with woody biennial stems, the flowers and fruits appearing with the year-old wood. The stems and leaf stalks are armed with slender, straight prickles which are dark and which contrast with the pale green of the new shoots. The leaves are pinnate, with three to seven broad, toothed leaflets which are sharply pointed, bright green above and often densely covered underneath with white hairs.

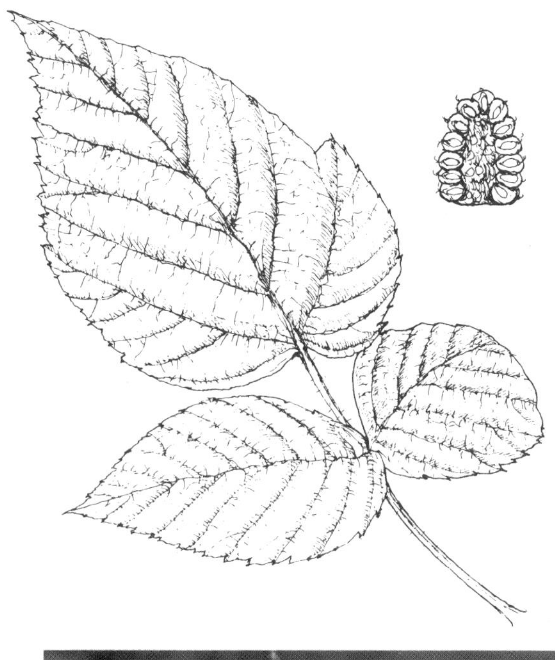

**Raspberry**

**White raspberry**

The leaf pedicel is channelled. The flowers are borne in loose, drooping clusters of one to ten, both at the tip of the stem and in the upper leaf axils. The sepals are lanceolate, pale green and hairy; the petals are white, erect and ovate, equalling the calyx in length. Raspberry flowers from May to August, although the late flowers may not bear fruit.

*Ecology and Distribution*
Raspberry is widespread and frequent in open woodland, on chalk downs, heaths, and roadsides throughout Britain, except Orkney and Shetland. It prefers drier, sandy or well drained soils, and grows with luxurious abundance in many hill areas of northern England and southern Scotland; here the canes are taller, lusher and with larger and more abundant fruit than those on the chalk downs of southern England. Raspberry grows commonly in Europe except in the far north and parts of the Mediterranean.

*History and Uses*
The raspberry has been used for centuries,both as a dessert fruit and in the making of jams, sweets and wine. The early herbalists knew it as Raspis, and Gerard refers to 'Raspis Bush'. The scientific name, *idaeus*, comes from Mont Ida in what is now north west Turkey, where the plant grows in profusion. The fruits contain pectic, citric and malic acids, the sugar fructose, and a volatile oil; raspberry syrup is used commercially as a permitted flavouring and colouring agent.

Raspberries are grown in gardens throughout the British Isles. In commercial production Britain leads the world, the largest areas being in south east Perth and Angus – the well known Carse of Gowrie.

# · *BRAMBLE* ·

***Rubus fructicosus* L. sensu lato**

The identification of brambles is complicated and fraught with difficulties for any one seeking to increase his knowledge of the group. The majority of brambles are capable of reproducing apomictically, being able to set seed without fertilization taking place; each bramble will be constant in form and of equal status to the others. Occasionally they will reproduce sexually with a male nucleus derived from pollen, thus giving rise to hybrids or to new unique apomictic forms. One of the most ubiquitous brambles, *Rubus ulmifolius*, is a sexual diploid, reproducing from male pollen and female ova, and it is inevitably a prolific hybridizer. Pollination is always necessary for the production of fruit.

Over 1000 different brambles have been recorded in Europe, and in Britain there are more than 290 recognized taxa, of which 202 are endemic.

One route sought out of this taxonomic morass was to group similar forms together under what A.Newton calls a 'senior' form. In many cases the senior form selected was a species which had been named historically early, while one of the 'juniors' in the group ultimately proved to be far more common and widespread. The forms tended to overlap, and some species would fit equally badly into more than one form, so Newton suggests that the method should be abandoned.

Criteria used in identifying brambles include:

1 The type of prickles and glands on the stem.
2 The shape, degree of toothing and hairiness of the terminal leaflet.
3 The structure of, and prickles on, the flowering branch.
4 The floral details.

Even this system can be misleading, for the leaf size, shape, texture and prickliness can vary greatly depending on the locality. In shaded sites rich in humus the bramble will be big and floppy, while plants growing in dry, sunny sites will be more prickly and have smaller leaves.

Species distribution will to some extent depend upon minimum winter temperature, so the number of possible species which can be found in any one locality decrease from south to north, ranging from sixty-two in south Devon, to five in Caithness.

*Rubus* species are very much plants of low ground, and do not ascend into the hills to any appreciable altitude, so that mountainous areas have a poor bramble flora. Some species such as *Rubus dasyphyllus* are almost ubiquitous, except for small areas where they are crowded out by more vigorous local species. Other species show a disjunct distribution due to human activity, such as along railway lines, while others have been spread by birds. Brambles flourish particularly well in oak-woods on rich loamy soils.

While acknowledging that each bramble species is an unique entity, the use of 'sections' can assist in the identification of an unknown bramble. The Subgenus *Rubus* can be divided into the following Sections:

SECTION SUBERECTI P.J.Mueller
A small group found in north and west Europe on heath, moorland borders, and upland woods on peaty soils. It includes *R.plicatus*, *R.nessensis* and *R.scissus*.
Characters – stems erect and smooth, strongly suckering, rarely rooting at the tip. Prickles subequal on angles of stems, stalked glands absent. Leaves usually green below. Sepals green externally, often with white margin. Fruit black or reddish-black.
SECTION CORYLIFOLIA J.Lindley
Group contains *R.tuberculatus*, *R.conjungens* and *R.sublustris*.
Characters – low arching, rooting, unridged stems with prickles all round and in some species stalked glands. Fruits may be pruinose.
SECTION SYLVATICI P.J.Mueller
Group contains *R.polyanthemus*, *R.lindleianus* and *R.nemoralis*.
Characters – stems arching, angled, often sparsely hairy and rooting apically. Prickles subequal, confined to the angles of stems. Stalked glands absent or few. Leaves green – leaves on upper part of stem may be grey and felted underneath. Sepals green felted or grey and woolly.
SECTION DISCOLORES P.J.Mueller
Group contains *R.ulmifolius* and *R.anglocandicans*.
Characters – stem often hairy, arching and rooting apically. Prickles subequal, confined to angles of stem. Stalked glands absent or few. All leaves grey-white tomentose underneath. Sepals grey-white tomentose and folded downwards.
SECTION SPRENGELIANI Focke
Group contains *R.sprengelii* and *R.lentiginosus*. The group is widespread in old woodlands and on heaths.
Characters – stems rather weak, procumbent or rooting at tip, slightly angled, maybe hairy and with a few stalked glands. Sepals grey, spreading or clasping the fruit.
SECTION APPENDICULATI (Genevier) Sudre
Group contains *R.vestitus*, *R.longithyrsiger* and *R.radula*.
Characters – stems strongly angled, arching, hairy

and rooting apically. Prickles unequal, scattered over stem surface. Stems and inflorescence with stalked glands. Inflorescence often compound. Sepals grey, tomentose, reflexed.

SECTION GLANDULOSI P.J.Mueller

This contains two well-marked groups, the *Hystrices and the Euglandulosi*. The former are often abundant and widespread in England, becoming scarcer northwards, and include one of the commonest brambles in Britain, *R.dasyphyllus*.

Characters – stems strong, arching, angled, with abundant prickles of various sizes, many gland tipped, and numerous stalked glands.

The *Euglandulosi* are relatively scarce in Britain, being found chiefly in ancient wooded areas in mid and south England. The most widespread (but thinly distributed over England and Wales) is *R.pedemontanus*.

Characters – stems weak, low arching, often pruinose, hardly angled, with dense rather weak armature of stout and slender prickles, many gland tipped, and long and short stalked glands.

A new approach developed by A.Newton has been to map the brambles as communities, each community consisting of a number of bramble species which consistently occur together. When this is done, six distinct regions appear, each with its own bramble florula. The North Sea florula is bounded to the west by the Pennines and contains brambles which commonly occur in Holland, Ger-

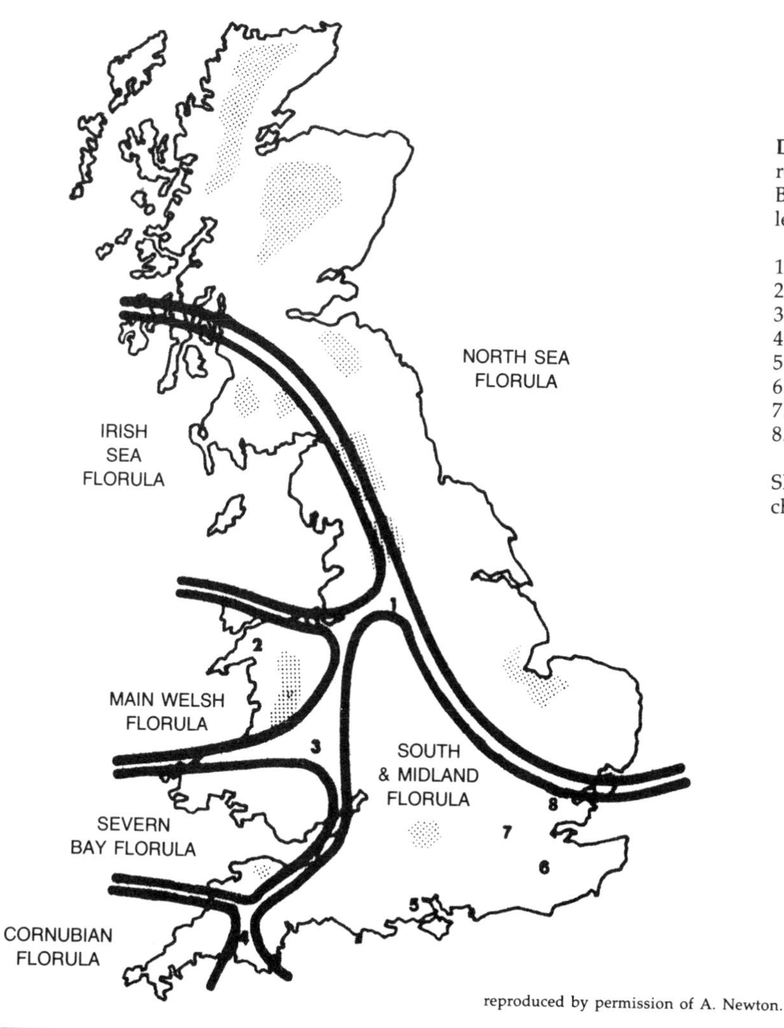


**Distribution of** *Rubus* florulas and regional endemic complexes in Great Britain. Main florulas identified by legend, regional complexes by number:

1. Sub-pennine
2. Padarn
3. Archenfield
4. South Devon
5. New Forest
6. Ashdown
7. Thames Valley
8. North Essex

Shaded areas denote high, wet or chalky ground with few or no brambles.

reproduced by permission of A. Newton.

many and Denmark. The South and Midlands florula contain brambles in common with those of north west France and Belgium, while the Cornubian florula lacks most of the continental brambles. To the west, the Severn Bay, Welsh and Irish Sea florulas each consist of a unique community of bramble species. There also exist eight regional complexes of brambles endemic to Britain, which overlap the boundaries of the main florulas and are pre-eminent in their own areas. Using this system it is possible to predict the main elements of the bramble flora of a given area, while always allowing for the occurrence of disjunct populations.

*Fruit*

The fruit of the bramble, the blackberry, is an almost globular drupe measuring 10-20mm in diameter, and composed of fifteen to eighty drupelets, each crowned with the shrivelled remnant of a style. Each drupelet, 3.5-4.5mm × 2.5-3.0mm, is black and shiny, with soft flesh and purple-staining juice. Within each is a single stone, 2.0-3.0mm × 1.5-1.7mm, brown with a rough, reticulate pattern on the surface. The ripe berry separates with the fleshy receptacle, leaving an almost flat surface, unlike the hollow in the centre of the

**Bramble (*Section Discolores*)**

raspberry left by its long conical receptacle. There is an enormous variation in size and texture between the fruits of the various bramble species, from small hard-textured blackberries with tiny drupelets to large, fat juicy fruits. *R.ulmifolius* has hard dense fruits which ripen late, but are sweet and last even into December. The size of the fruits will also vary greatly from year to year on the same bush, depending on the amount of rainfall in mid and late summer, drought years being accompanied by small, shrivelled fruits. The flavour of the different species also shows considerable variation, but is always sweet and pleasant. Blackberries are at their best from the end of July until the end of September, after which the flavour becomes insipid. The berry at the tip of the inflorescence always ripens first, and is often the largest.

*Leaf and Flower*

The various bramble species are very variable, prickly, climbing or prostrate, half-evergreen perennials, with woody stems up to 4m long. The tips of the stems of some species will root in autumn and will carpet the ground, forming dense thickets. The stems vary greatly in size and armament from delicate greenish stems with fine, straight prickles, to massive angular stems, deep purple in colour, armed with equally massive hooked prickles. The leaves are pinnate, with three to five broadly toothed leaflets, the basal pair usually stalked, with prickles down the leaf stalks and on the underside of the larger veins. The flowers, 15-30mm across, are borne in panicles on the end of the previous year's stems and have five white or pink petals – often crinkled – and pointed green sepals which usually turn down as the fruit ripens. Bramble flowers appear from May to September and attract many butterflies and other insects to their ample supplies of nectar. They are sweetly scented. In autumn the leaves turn an attractive deep reddish purple.

*Ecology and Distribution*

Brambles are widespread and abundant in woods, hedges and waste places, and on heaths and cliffs throughout Britain. They form dense thickets, and carpet the floor of neglected woodlands. Even on bare shingle, the prostrate stems run in all directions, each many metres long. Brambles support a vast and varied animal population, acting as a prickly shelter for small mammals and birds, their blossoms offering nectar to a host of insects and butterflies and their fruits being of especial value to birds on migration. Seventeen species of birds have been recorded eating blackberries in the period from August to December, but especially in September and October. Blackbirds are the commonest predators; other species include ring ousels, songthrush, starling, bullfinch, greenfinch, blue tit, great tit, whitethroat and lesser whitethroat. It is quite disconcerting to be faced with a whitethroat or greenfinch with its face stained purple with blackberry juice, and many birdwatchers must have momentarily wondered if they were seeing some rare and exotic accidental.

*History and Uses*

Blackberries have been eaten with relish by man from earliest times; a well-preserved Neolithic body recovered from the Essex clay had blackberry pips in the stomach content. They are still used in jams and preserves, syrups, vinegars and wines, and in the classic dessert of blackberry and apple pie. Culpeper advised that 'the berries of the flowers are a powerful remedy against the poison of the most venemous serpents; as well drank as outwardly applied, helpeth the sores of the fundament, and the piles'. Blackberry vinegar is especially favoured for the treatment of coughs and colds. Once October is come blackberries in parts of southern England are called 'the Devil's fruit' and left alone, probably because by that time they have become squashy and flavourless.

Two other bramble species have become so well established as naturalized aliens that they are worth mentioning. *Rubus spectabilis* can be quickly distinguished by its chestnut bark and bright, emerald green leaves. The petals curl up like a pink Chinese lantern, and the fruit has an incredible luminous colour, somewhere between marigold and salmon pink. *R.spectabilis* is well established in the west of Scotland, particularly in Arran, where it is known as Arran Raspberry. Allan Stirling has described it as filling a small valley in Caithness. In 1983 it was recorded in Co Armagh, and in 1985 it was found in three locations in mid-Cork. *Rubus procerus* is widely cultivated in gardens as 'Himalayan Giant'. It belongs to Section *Discolores*, is a very large and vigorous grower, and has leaves with silver reverses and very large fruit. It is well established in Scotland, and was recently recorded in Warwickshire.

**Bramble (*Section Sprengeliani*)**

# · DEWBERRY ·

**Rubus caesius L.**

*Fruit*
Although similar in many respect to the brambles, dewberry stands apart and is easy to identify. The fruits are the most beautiful of all the *Rubi*. Each rounded dewberry, 12-15mm in diameter, is composed of a relatively small number of up to twenty-five very large, fleshy drupelets, dark blue with a dense grape-like bloom on the surface. Close examination of the fruit reveals that each drupelet is almost triangular, 5mm long and 4mm broad, with a persistent whiskery style protruding from its broad outer surface, and that the bloom is quickly lost with handling.

Dewberries are very juicy, although not really sweet. The juice is watery and bright red, unlike the purple-staining juice of the bramble. Within each drupelet is a single pale brown, rough-coated stone 3-3.5mm long, curved like a comma. The seed of the bramble lacks this curved apex.

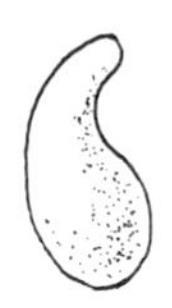

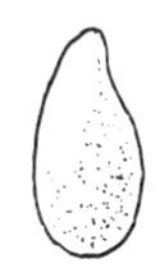

Dewberries are ripe from late July, several weeks earlier than blackberries, and last until the end of September. The best time to find them is late August.

*Leaf and Flower*
Dewberry is less robust and prickly than the brambles, and is often prostrate. The stems are pruinose when immature, rounded and smooth, or armed with fine, straight prickles. The leaves are always trifoliate, wrinkled, paler green than bramble leaves and hairy on the upper surface. In early autumn they turn a brilliant russet-red, and in winter are deciduous. The flowers, 20-25mm in diameter, are larger than most bramble flowers and are always white, crumpled and round, blooming from June until September. They are borne in small branched clusters. The sepals are grey-green and with white margins and long points, clasping the developing fruit and scarcely down-turned except when the fruit is very large.

*Ecology and Distribution*
Dewberry has the remarkable genetic capability of acting as pollen parent to form sterile, or occasionally fertile, hybrids with other *Rubi*, while remaining a virtually obligate apomict as a seed parent. It is the putative parent of the group of brambles known as *Corylifolii*, which consist of stable, fixed microspecies, often of widespread occurrence. It is widespread and frequent in bushy or dry, grassy places, in fens and dune slacks, and in woods on calcareous clay. It tolerates damper places than those where other brambles flourish, and often

grows on limestone and chalk. It is commonest near the coast and in the south and east of England, with a sparse distribution in most of lowland Wales and Ireland. In Scotland it is confined mainly to the south east and the Solway coast. In many downland areas it is best sought in the base of old chalk workings or on broken banks, where it flourishes among the chalk rubble.

It is found in most of Europe except some of the Mediterranean islands, such as Corsica, Sardinia and Crete, and it is absent from the islands of the far north.

*History and Uses*

Shakespeare mentions them in *A Midsummer Night's Dream*, where he has Titania instructing her fairies to be kind to Bottom –

'Feede him with Apricocks and Dewberries,
With purple Grapes, greene Figs and Mulberries'.

Culpeper in his Herbal of 1652 named the Dewberry the 'Gooseberry Bush, called in Sussex Dewberry Bush'.

Dewberries may lack the strong flavour of blackberries, but they are tangy and delicious, both raw and cooked.

**Dewberry**

# · THE ROSES ·

The Genus *Rosa* presents a bewildering spectrum of species and hybrids which at first sight may daunt the inexperienced botanist. The field rose (*Rosa arvensis*) and the burnet rose (*R.pimpinellifolia*) are balanced diploids or tetraploids (having two or four sets of chromosomes). In the cell divisions accompanying sexual reproduction the sets of chromosomes divide in a regular and ordered fashion, the male and female components being contributed in equal proportions. These species are consistent in form and readily identifiable, but most of the others are unbalanced polyploids, mainly pentaploids with five sets of chromosomes, and they behave during sexual reproduction and hybridization in a manner unique to the Genus *Rosa*.

For example, *Rosa canina*, one of the Dog Roses, is pentaploid, having thirty-five chromosomes in five sets of seven each. During meiosis two sets form a pair, and the remaining three sets stay unpaired. When male pollen is formed the three sets are reabsorbed and lost, the remaining paired set contributing one set to each pollen grain, which will thus contain seven chromosomes.

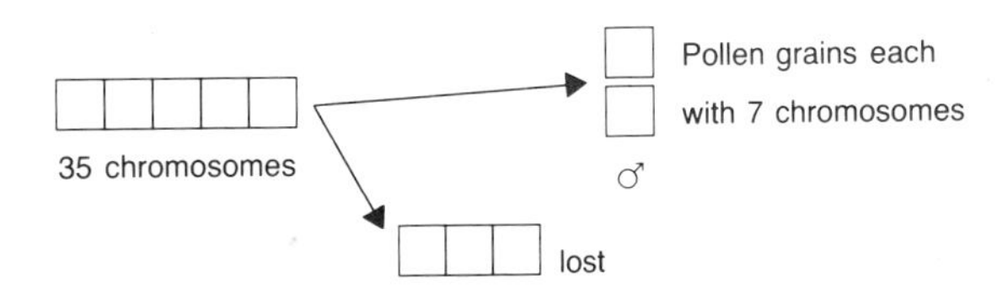

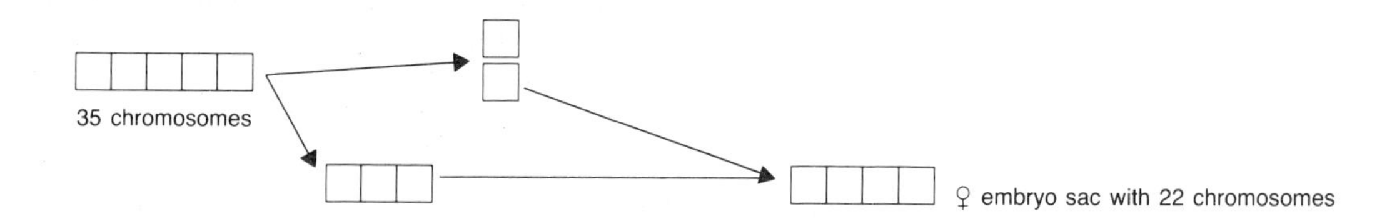


When the female embryo sac is formed, the three unpaired sets link with one of the paired sets, to give a set of four which will contain twenty-eight chromosomes.

At fertilization the fusion of the single set from the pollen grain and the four sets from the embryo sac restores the full pentaploid complement of thirty-five chromosomes, and the resulting plant will thus be four-fifths female derived and one-fifth male derived. The progeny of hybridization between roses will therefore show a strong resemblance to the species which contributed the female element and the hybrids will range from completely infertile to relatively highly fertile. The pattern of hybrids within an area will also be influenced by the basic availability of different species, so that in Ireland for example the possible hybrids can only be between those roses which have a western distribution pattern within the British Isles.

In the identification of roses the following characters must all be examined:

**1** Habit – the manner in which the plant grows.
**2** Leaf shape and type of marginal serration.
**3** The type of armature and glands on the peduncle.
**4** The presence of hairs on the leaves.
**5** The colour, size and scent of the flowers.
**6** The characteristics of the fruit, its colour, shape, texture, presence of sepals on the ripe fruit and the proportion of the stylar aperture at its apex.

Mature fruits are necessary for proper identification of roses, and care must be taken, since the characters may vary from one part to another on the same bush.

In this book, in order to simplify the approach to identification the roses which show a marked affinity to one another have been grouped together. Within these groups certain species have been selected for illustration, but they stand as individuals and not as examples of an aggregate. Some attempt has also been made to illustrate species common to an area of the British Isles, where there is a marked difference in the pattern of species distribution from north to south.

**1** Field Rose (*Rosa arvensis*).
**2** Burnet Rose (*Rosa pimpinellifolia*).
**3** Stylous Rose (*R.stylosa*).
**4** Dog Roses
  Northern (*R.coriifolia* and *R.afzeliana*).
  Southern (*R.canina* and *R.obtusifolia*).
**5** Downy Roses
  Northern (*R.mollis*)
  Intermediate (*R.sherardii*).
  Southern (*R.tomentosa*).
**6** Sweet Briars (*R.rubiginosa* and *R.micrantha*).

Many roses will be found which cannot be fitted neatly into these compartments. In many cases – unless one is a highly experienced rhodologist – it may not be possible to decide if an individual bush is a recognized species or a hybrid.

The fruit of the rose is called a hip, and is formed by the fleshy, urn-shaped receptacle. At the top of the hip is a round disc-shaped structure bearing the remnants of the numerous perigynous stamens, with the stylar opening at its centre. The relative size of this opening is an important diagnostic feature, and can most easily be observed in the ripe fruit. With a little dexterity, the stylar remnants can be pulled out, leaving the orifice clear and easily measured. The carpels are formed on the inner wall of the receptacle, and the seeds are achenes. The rest of the hip is packed with a mass of brittle, silvery hairs – these are used for making the itching powder beloved of practical jokers.

# · FIELD ROSE ·

**Rosa arvensis Hudson**

*Fruit*

The fruit of field rose is an ovoid hip, smooth and dark red, 10-16mm long and 5-6mm wide, with the long, pointed, simple sepals deflexed and falling from many of the hips as they ripen. The stylar remnants protrude like a black matchhead from the apex of the hip. The slightly glandular pedicel is about 13mm long. A section cut through the hip shows the rounded shoulders of the stylar aperture, and the pale brown, markedly triquetrous seeds, 5 × 2.5mm

The hips are at their finest in October.

*Leaf and Flower*

Field rose is a weak, clambering, deciduous shrub, growing to 1m unless supported, forming brakes of arching, graceful stems. The stems may be purplish, and carry narrow-based, hooked prickles which are red when young. The pinnate leaves have two or three pairs of simple serrate leaflets, with large, shining, nearly entire stipules at the base. The leaflets are well spaced and hairless, glaucous beneath and aglandular. The penduncles are long, usually glandular, and carry clusters of one to six cup-shaped flowers 25-50mm in diameter, which are always creamy white in colour. They are smaller than the flowers of Dog Roses, and appear a little

**Field rose**

later. The styles form a slender column projecting from a flat disc as long as the numerous yellow stamens. Field rose flowers from June until August.

*Ecology and Distribution*
Field rose is fairly tolerant of shade and poor drainage, and is widespread and not uncommon in open woods, thickets and hedgerows in England, Wales and south east Ireland. In northern England as far north as Yorkshire it is local; in Scotland it is rare and doubtfully native; near Aberdeen it is naturalized. Elsewhere plants have proved to be of hybrid origin.

Field rose is widespread in south, west and central Europe.

# · BURNET ROSE ·

***Rosa pimpinellifolia*** **L.**

*Fruit*
The hips of burnet rose are easy to identify, the mature fruits being purplish black, flattened spherical in shape, 10-15mm in diameter and 8-10mm high. They are smooth and shiny, crowned with erect, prominent and pointed sepals, which often

**Burnet rose**

enclose a shrivelled mass of stamens. The wall of the hip is dark red and hard, enclosing the large seeds, blunt oval in shape, 4mm long and shiny yellow. The pedicel is often dark red and smooth, although the peduncle may be bristly. The hips are ripe in late August.

*Leaf and Flower*
Burnet rose is a creeping, deciduous shrub up to 1m high, suckering readily to form extensive patches, especially where the soil is light and sandy. The stems are densely covered with long and short straight spines and fine bristles. The pinnate leaves bear five to eleven small rounded leaflets, simple serrate, and glabrous on both surfaces. The flowers are solitary, without bracts, and normally of an ivory white colour, although deep pink-centred flowers have been seen in sand dunes on the Isle of Man and in west Sutherland. They have a delicious, sweet scent. The sepals are narrow and undivided, persisting and erect on the ripe fruit, and the styles are woolly. Burnet rose flowers from May to July.

*Ecology and Distribution*
Burnet rose is widespread on dry stabilized sand dunes, heaths, downs and limestone pavement, especially near the sea, and on calcareous mountain ledges up to 500m. It is rather uncommon in south east England, but occurs locally north to Caithness and the Outer Hebrides. It is particularly luxuriant in Ireland, growing in the limestone pavements of the Burren in Co Clare and Galway. It is widespread in Europe except in the north east, most of Fennoscandia, the extreme south west, and many islands.

# · *STYLOUS ROSE* ·

**_Rosa stylosa_ Desv.**

*Fruit*
The hips of stylous rose are long, smooth and elegant, 10-15mm long and orange red. They are borne in long-stalked clusters of one to eight on pedicels about 27mm long, sparsely covered with glandular hairs. The sepals usually fall from the ripening fruits and are markedly pinnate, unlike the simple sepals of *R.arvensis*, and more closely resembling the sepals of *R.canina*. The stylar remnants are prominent, projecting from a conical disc. The seeds are smooth and yellow, slightly angled, 4.5-6mm × 3-3.5mm. The hips are ripe from August until the end of October.

*Leaf and Flower*
Stylous rose is a tall spreading shrub 1-4m high, with stout, strongly arching stems armed with stout, hooked, straw-coloured prickles. The pinnate leaves bear three to seven well spaced leaflets, long and elegantly pointed, dark green and simple serrate. The leaf surfaces are smooth, with a few glandular hairs on the ribs and leaf stalks, and the undivided stipules are edged with glandular hairs.

The large flowers, 30-50mm across, are white or pale pink, (*R.arvensis* is always white), and the sepals are markedly pinnate, folding down when the petals fall and the fruit starts to ripen. The smooth styles protrude from the top of the conical disc in a fountain shape, not a column. Flowering takes place in June and July.

*Ecology and Distribution*
Stylous rose grows in sheltered hedges on basic soils. In England, it has a mainly southern distribution, north to Suffolk, Lancashire and Worcestershire. In Wales it is more frequent in the south. In Ireland it is recorded from Cork and South Tipperary, but may prove to be more widespread. In Europe it is recorded locally from West Germany to southern Spain, and east to Bulgaria.

**Stylous rose**

# · *The Dog Roses* ·

The name 'Dog' probably derives from the Old French *dague* (hence dagger), and refers to the fierceness of the thorns. Dog Roses are robust bushes with leaves glabrous or only sparsely hairy (unlike the Downy Roses) and lacking the apple-scented glands of the Sweet Briars. They are a large group showing many different characteristics and great variability, even within a species; they also hybridize freely.

# NORTHERN DOG ROSES

**_Rosa coriifolia_ Fries** AND **_R.afzeliana_ Fries**

## *Rosa coriifolia*

*Fruit*
The ripe hips are rounded, red and shiny, 20-25mm in diameter, crowned with erect, slightly spreading sepals. The pedicels are smooth and short, so that they are hidden in the broad stipules. The seeds are 5 × 3mm, angled and yellow, and the hips ripen in August to October.

*Leaf and Flower*
The bushes grow to 3m, with arching stems armed with small hooked prickles. The leaves have five to seven leaflets which are doubly serrate, pubescent above and beneath. The flowers are 40-50mm in diameter, varying in colour from white to red, and the sepals are markedly pinnate. The styles form a domed woolly head and the stylar orifice is one-third the width of the disc.

*Ecology and Distribution*
This species is most common in northern England and Scotland.

*Rosa afzeliana*

*Fruit*
The large hips are rounded, red and shiny, crowned with erect, slightly spreading, pinnate sepals which persist. The pedicels are short and smooth. The fruits tend to ripen earlier than those of *R.coriifolia*.

**Rosa afzeliana**

*Leaf and Flower*
The foliage is glaucous and the flowers are deep pink. The styles form a rounded, woolly mass over the stylar orifice, which is one-third the width of the disc.

*Ecology and Distribution*
This is the commonest wild rose in central and northern Scotland, occurring widely in Wales where it is not common, and very rarely in southern England.

# SOUTHERN DOG ROSES

**Rosa canina L.** AND **R.obtusifolia Desv.**

*Rosa canina*

*Fruit*
The hips of this Dog Rose are bright red, smooth and shiny, about 15mm long and 9mm wide. They are slightly egg-shaped, but show great variability even between hips on the same bush, so that the widest part may be above or below the midline. The pedicels are short, about 8mm, and smooth. The pinnate sepals reflex and lie against the immature fruit, but are shed before the hips are ripe in September and October. The seeds are markedly triquetrous 4.5 × 3mm.

**Rosa canina**

*Leaf and Flower*
This Dog Rose is a robust, deciduous bush 1.5-3.5m high, with strong arching stems and stout, hooked prickles dilated at the base. The pinnate leaves have five to seven simple serrate leaflets which are

**Rosa canina**

smooth on both surfaces, with glandular hairs and small prickles along the ribs and stalks of the leaflets, and glandular hairs along the edges of the stipules. The leaves are shed in the autumn earlier than the leaves of any other species, leaving the bare bush crowned with red hips which persist well into the winter. The flowers are large, 40-50mm across, pink or white, appearing from May to July, in clusters of one to four. They are fragrant, with a mass of yellow stamens, and smooth styles in a loose cone. The disc orifice is one-fifth its width.

*Ecology and Distribution*
This is the commonest Dog Rose in southern England, less common in Scotland. It is widespread in thickets and hedges, but though tolerant of widely differing conditions it evinces a dislike for elm-dominated hedges.

## *Rosa obtusifolia*

*Fruit*
The hips are squat and globular, 9-11mm in diameter, smooth, red and shiny, much rounder than the hips of *R.canina*. Pedicel length is about 12mm, and the seeds are rough, triquetrous 4.5 × 2mm. The sepals are shed before the fruits are ripe.

*Leaf and Flower*
The growth habit is spreading, with arching stems armed with ferocious, hooked prickles. The five to seven leaflets are broad and rounded, double serrate, with hairs and glands on the underside of the veins and midribs. The flowers are 35-45mm in diameter. The styles from a flat dome, and the stylar orifice is one-fifth the width of the disc.

*Ecology and Distribution*
This Dog Rose is well distributed in scrubland in the south and east of England but is never common. Its range extends north to Northumberland and Cheshire.

**Rosa obtusifolia**

# · The Downy Roses ·

The three Downy Roses selected here show an interesting gradation in characteristics, and are good examples of the species to be expected in the south, west and north of the British Isles. The following key to identification has been supplied by the Rev G.G.Graham.

| | *R.tomentosa* | *R.sherardii* | *R.mollis* |
|---|---|---|---|
| Sepals on ripening fruit | Falling early |  |  |
| Stylar aperture | ⅕-¼ disc | ⅓ disc | ½ disc |
| Styles | Often only hispid | Woolly head | Large woolly head |
| Pendunces | Long, glandular | Medium length, glandular | Short, glandular |
| Prickles | Strongly hooked | Slender, slightly curved | Straight and slender |
| Distribution | Mainly southern | More in north, but also in Wales and Cornwall | Northern |

## *Rosa tomentosa* Smith

*Fruit*
The hips are dark red and shiny, 18mm long and 8mm wide, broadest below the middle. The surface of the hip usually bears a few scattered, glandular hairs, and the long pedicel (22mm) is well supplied all round with rather stiff glandular hairs. The seeds are rough and irregular, 6 × 2.5mm. The sepals fall from the hips before they are fully ripe.

*Leaf and Flower*
This is a tall rose with arching stems and stout, hooked prickles. There are five to nine fairly broad leaflets, double serrate, the upper surface dull dark green and pubescent, the under surface densely covered with fine soft hairs. There are glandular hairs along the edges of the leaflets and stipules. The flowers are 45-55mm across, varying in colour

***Rosa tomentosa***

from white to red, and the sepals are pinnate with long glandular hairs on the outer surface and dense woolly hairs within. The flowers open in early June and July.

*Ecology and Distribution*
*R.tomentosa* is widespread in England, Wales and Ireland, becoming rare north of Durham, and not known for certain in Scotland. It is a local plant of scrub and wood margins on both acid and calcareous soils.

***Rosa tomentosa***

## *Rosa sherardii* Davies

*Fruit*
The hips are large, dark and somewhat bristly, ripening early from August onwards. The sepals remain on the ripe fruit, suberect at an angle of 45°. They are not large and have rather small lateral pinnae.

*Leaf and Flower*
*R.sherardii* is a compact bush armed with slender, slightly curved prickles. The young stems and

**Rosa sherardii**

leaves are often pruinose, with a blue-green sheen. The double serrate leaflets are soft and grey green, with fine hairs on both surfaces, and many strongly aromatic glands below. The flowers are dark pink to red, with a woolly head of styles and a stylar opening one-third the width of the disc.

*Ecology and Distribution*
*R.sherardii* is widespread and very common in hedges in the north of England and Scotland, but also has a more western range extending into all of Wales and south west into Cornwall. In Ireland it is recorded from Kerry, Co Cork, and Waterford, and is probably more frequent than records there suggest. In 1983 it was recorded in Suffolk for the first time.

## *Rosa mollis* Smith

*Fruit*
No one travelling the country roads of Scotland in September and October can fail to see the spectacular display of huge, dark red hips of this and *R.sherardii*. The hips, borne on short glandular pedicels, are 20 × 15mm, irregular oval and shiny, with scattered bristly, glandular hairs on the surface. They are crowned by the erect, large and prominent sepals, often longer than the hip, and reminding one of the twisting tentacles of a squid.

The seeds are triquetrous, smooth and yellow, 5 × 3mm. One variety var.*pomifera* (Herrn.) Desv. has spectacular, large, apple-shaped fruits.

*Leaf and Flower*
*R.mollis* is an erect bush armed with straight prickles. The leaflets are soft and rounded, double serrate, finely downy above and downy glandular below, especially on the ribs and edges of the leaflets. The flowers are large, 30-50mm across, and often deep pink or red. The simple sepals are large and very glandular, often expanded at the tip. The styles form a large woolly head, and the stylar aperture is half the width of the disc.

*Ecology and Distribution*
*R.mollis* is widespread and common in hedges and roadsides throughout Scotland and the north of England. Its range extends south to Derby, Hereford and Glamorgan. To the west it is local in Ireland.

***Rosa mollis***

# · *The Sweet Briars* ·

The Sweet Briars are distinguished by sweet, apple-scented glands on the foliage, so that by crushing and smelling a leaf they are easily identified. Two of the more common species are described here. Sweet Briars are particularly susceptible to attacks by rose gall wasps, forming fragrant Robin's Pincushions.

## *Rosa rubiginosa* L.

*Fruit*
The hips are 10-15mm in diameter, bright scarlet and shiny, varying in shape from ovoid to egg-shaped; there may be scattered glandular hairs on them. The erect sepals persist on the hips as they ripen. The seeds are triquetrous, 5 × 3mm.

*Leaf and Flower*
*R.rubiginosa* is a small erect shrub of 2m, armed with stout, unequal, hooked prickles. The pinnate leaves have five to nine small rounded leaflets, double serrate, usually smooth, and well endowed with sticky brown, apple scented glands on the lower surface. The pink flowers, 16-30mm across, are carried in clusters of one to four on short pedicels which are densely covered in glandular hairs and fine prickles, and the styles are hairy.

*Ecology and Distribution*
*R.rubiginosa* is widespread and locally common in England and Wales, mainly in scrub on calcareous soils. It is local in Ireland and rare in Scotland, being absent altogether from Orkney and Shetland. In England it is most common south east of a line from the Wash to Dorset.

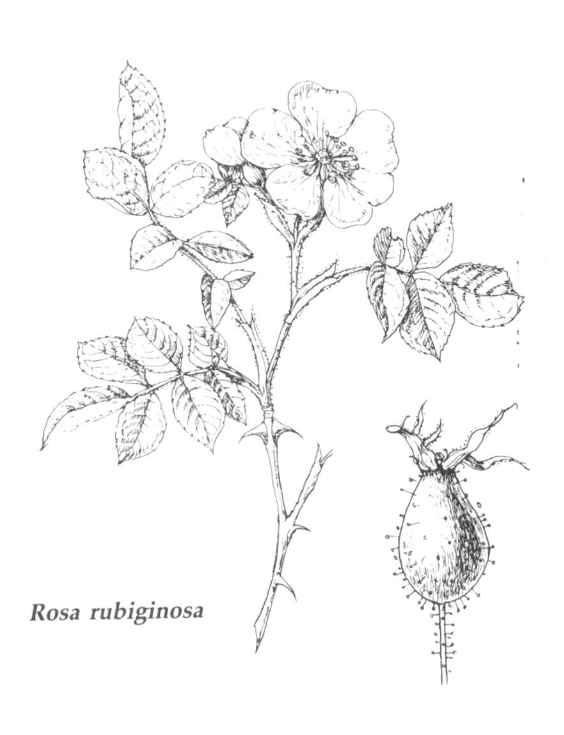

**Rosa rubiginosa**

**Rosa rubiginosa**

## *Rosa micrantha* Borrer ex Smith

*Fruit*
The hips are smaller than those of *R.rubiginosa*, 8-10mm long, mainly smooth or with a few glandular hairs, and often flask-shaped, with a marked constriction at the top below the disc. The pedicels are longer than those of *R.rubiginosa*. The sepals fold back and fall early, before the fruit ripens.

*Leaf and Flower*
*R.micrantha* has arching stems up to 3.5m, armed with large uniform prickles, and leaflets similar to those of *R.rubiginosa* but larger. The flowers are smaller, and the styles are smooth.

*Ecology and Distribution*
It is widespread in calcareous scrub, heaths and pastures in England and Wales, and appears to be absent from Scotland. In Ireland it occurs mainly in the south west, but in most of this range it seems to be replaced by hybrids.

*History and Uses (refers to all the rose species)*
Rose flower petals have been used in the past to flavour jams, wines and sweets. The fruits have been used in desserts since at least the Middle Ages although all the writers mention the tedious job of first removing the seeds and short brittle hairs from within the hips.

Culpeper wrote: 'The pulp of the hips has a grateful accidity . . . good for coughs and spitting of blood. The hips are . . . fitly given to consumptive persons'. He also suggested using the larvae of the gall wasp (*Rhodites rosae*) which cause the fluffy-looking Robin's Pincushions on Sweet Briars and Dog Roses, as a vermifuge in children.

Robert May (1671) described Rose Hip Soup, a Tarte of Hips, and syrup, jam and wine made from rose hips.

Deakin (1871): 'The fruit, when ripe, is pleasantly acid, and is used for making a kind of preserve, which is very useful in sore throats and is often used on the Continent as dessert and in confectionery'. Dr Losch, writing in the 19th Century, described the use of Dog Rose hips (*Fructus Cynosbati*) for liver and kidney disorders, and a conserve of wild rose hips 'Conserve de Cynorrhodons' to treat diarrhoea.

The use of rose hips played a part in the 1939-45 War. In Britain no citrus fruit was available, so rose hips became an alternative source of vitamin C. The ripe, fresh hips contain 0.1-1.0% ascorbic acid, with vitamin A, thiamine, riboflavine and nicotinic acid, three to four times as much as blackcurrant juice, and ten times as much as orange juice. In 1941 the Ministry of Health promoted a scheme to gather rose hips, and 120 tons were collected, this quantity increasing to an annual crop of 450 tons. The rose-hip syrup thus made was distributed to children and expectant mothers, a delicious supplement to a wartime diet, and rose hips are still gathered by voluntary collectors, especially in the north of England.

Bird predation of rose hips is dependant on the severity of the winter weather, and blackbirds are the most frequently recorded species eating rose hips.

***Rosa micrantha***

# · WILD STRAWBERRY ·

***Fragaria vesca* L.**

*Fruit*

The strawberry is formed by the much enlarged fleshy receptacle. The true fruits are the pips: small dry achenes, which are set over the entire surface of the strawberry and, in this species, protrude from it. The fruit of the wild strawberry droops when ripe, from June to August, sometimes later. It is red, sweet and juicy, with a delicious taste and fragrance which contrives to be both more delicate and at the same time more pungent than that of the cultivated forms. The fruits are small, often less than 10 mm in diameter, and vary greatly in shape from flattened spherical to pointed oval. Plants with white fruits, which occur rarely, have yellow pips on the surface and a particularly sweet flavour.

*Leaf and Flower*

Some difficulty may be encountered in the identification of the strawberry growing wild, especially if there is no ripe fruit. It is a low-growing perennial, with long arching runners which root at the internodes and form new plants. The name 'Straw' derives not from the custom of putting straw under garden strawberries to keep the fruits clear of the soil, but from the old verb 'to strew', relating to the tangle of stems and runners spread on the ground. The trefoil leaves are long-stalked. The pointed leaflets have deeply set veins and coarse teeth. The leaves are thin, light green and slightly glossy above, with fine silky hairs sparsely scattered on the leaf surface. The terminal tooth of each leaflet is

as long or longer than those on either side. The leaf stalks are densely covered with spreading white hairs, and the flower stalks, especially the uppermost pedicels, bear a mixture of appressed and ascending hairs. The flowers, 12-18mm in diameter, are carried in upright clusters of one to four on short stalks. The five petals are white, longer than the sepals, with no gaps between them and no notch at the outer edge. The green sepals are spreading or folded back as the fruit ripens. Wild strawberry usually flowers from mid-April to the end of June, but flowers may still be formed as late as October, although too late in the year to form fruit.

*Ecology and Distribution*
Wild strawberry is widespread and common in open woods and on grassy banks throughout the

**Wild strawberry**

British Isles. It is less common in the far north, and does not grow on hills over 700m, showing a preference for lime-rich or base-rich soils, well drained and in full sun. It is also widespread throughout most of Europe.

*History and Uses*
The first reference to strawberries is in a Saxon plant list of the 10th Century. In 1265 'Straberie' is mentioned in the household roll of the Countess of Leicester. Wild strawberries had been gathered and eaten raw from earliest times, and were widely used in sweets and conserves. One particularly delicious way to serve them is with fresh cream-cheese beaten with granulated sugar.

Gerard had many uses for strawberries. 'The leaves boyled and applied in manner of a pultis, taketh away the burning heate in wounds: the decoction thereof strengtheneth the gummes and fastneth the teeth. The ripe Strawberries quench thirst, and take away, if they be often used, the rednesse and heate of the face'. Strawberries have also been used in the treatment of gout. It is said that if the juice is held in the mouth for five minutes it will remove staining of the teeth, which must then be rinsed with warm water and a pinch of bicarbonate of soda. A cut strawberry rubbed on the face after washing will sooth the effects of slight sunburn.

Blackbirds are recorded frequently taking wild strawberries.

The 'Plymouth Strawberry' is a strange little monstrous form of *F.vesca*, which has a long history. It was mentioned both by Gerard and in Ray's 'Synopsis', and John Tradescant saw it growing in a garden in Plymouth early in the 17th Century. Below the fruit there is a double ruff of small-toothed leaves, and although the base of the fruit reddens when ripe each achene is drawn out into a green spike about 3mm long.

# · *GARDEN STRAWBERRY* ·

**Fragaria x ananassa Duchesne**
**(also Hautbois Strawberry *F.moschata* Duchesne)**

*Fruit*
The fruit of Garden strawberry varies greatly in form, depending upon the cultivars from which the particular plant has arisen. The fruits are much larger than those of *F.vesca* – 30mm or more in diameter, with the achenes sunk into the flesh of the receptacle. White fruits occur uncommonly. The Hautbois strawberry has a purplish red, oblong fruit, 10-25mm long, with no achenes at the base of the fruit, which may also be green tinged. Both species are ripe from late June to the end of July.

*Leaf and Flower*
Garden strawberry is bigger and stouter in all its parts than wild strawberry. The leaflets are more or less glabrous above, and the terminal lobe is rounded. The presence of hairs on the leaves and pedicels is not a reliable feature, and depends upon the age of the plant. The flowers are larger than those of *F.vesca*, 20-35mm across, and less erect. The calyx shape is not reliable as a means of identification, varying from tightly clasping the fruit to fully reflexed. The Hautbois strawberry seldom produces runners, and has dull green, deeply veined leaves which are heavily pubescent on both surfaces and relatively broader and less tapered than the leaves of the other two strawberries. The flower stalks are longer than the leaves, covered with spreading and reflexed hairs, and the flowers are smaller, 15-25mm across. Almost all naturalized *F.moschata* have functionally dioecious flowers. Both species flower from May to July.

The fruits of the three strawberries make their identification easy, but in their absence *F.x ananassa* can be distinguished from *F.moschata* by being less hairy, and from *F.vesca* by its larger flowers, thicker leaves and stouter petioles and runners.

Garden strawberry

*Ecology and Distribution*

Both strawberries grow in waste places, especially warm sunny banks on light or calcareous soils. *F.x ananassa* shows such a strong predilection for the sunny banks of railway lines that it well merits the name 'Railway Strawberry'. It is an introduced plant, and although widely distributed it is much less common in Ireland and the north west of Scotland, most of the records coming from the warmer, drier south east corner of England. It is widely cultivated and naturalized throughout Europe. Recent studies indicate that *F.moschata* may be native in Holland.

*History and Uses*

Most of the large fruited cultivars grown in Britain stem from hybrids of *Fragaria virginianum* and *F.chiloensis* which were introduced into Europe from Virginia in America in the 17th Century, probably in 1629.

Apart from their consumption as dessert, strawberries are widely used in jams, conserves and liqueurs, and as flavouring for a multitude of products, with numerous hybrids developed for commercial growing on a large scale.

# · PLYMOUTH PEAR ·

***Pyrus cordata*** **Desvaux**

*Fruit*
The fruit of the pear is a pome, formed as in rose hips by the wall of the receptacle, with, in this case, the extended bases of the carpels fused with it. The fruit of Plymouth pear is small, 8-18mm long, somewhat rounded and hard, carried on a pedicel 25mm long. The surface is shiny, red tinged and covered with lenticels. The apex of the ripe fruit is not crowned with the brown sepal remnants, as in the wild pear, since they are shed during development of the fruit. The flesh of the pear is tough and silicaceous, grating when cut. The fruit is four or five celled with two seeds in each cell, but not all of these mature, so that most fruits contain one to five pips like apple pips, 4 × 1.5mm, with a finely striate brown coat and a sharply pointed base. The fruit is mature in September and October, but in Britain it is sparsely produced.

*Leaf and Flower*
Plymouth pear is a deciduous, spreading shrub, normally 3-4m high, but occasionally reaching 8m. The lower branches are often spiny and the young twigs are purplish. The leaves are small, broadly ovate, and nearly smooth, with neat regular serrations on the margins. The flowers open from late May to mid-June. They are smaller than the flowers of wild pear, 16-20mm across, in a simple corymb, the outer flowers on longer stalks so that the inflorescence appears level-topped. The petals are pink, especially on the outside. The calyx is long and funnel-shaped, not constricted below, and covered with brown pubescence; the densely woolly sepals are deciduous, falling before the fruit ripens.

**Plymouth pear**

*Ecology and Distribution*
Plymouth pear is an extremely rare plant in England, known only from hedgerows at Egg Buckland near Plymouth in Devon. In Europe it grows in the west, from central Portugal, Spain and France north to south west England. It is found in hedges and on the edges of woods, usually on schistose soils and near the sea.

*History*
Plymouth pear has long been known at Egg Buckland. The hedge where it grew was cut in 1958, after which the bushes grew well but did not flower. In the late summer of 1966 some of the hedge was destroyed when a new main was laid, and the rest was threatened by levelling. Urgent steps were taken to preserve the bushes that remained, cuttings being taken with the aim of propagating and re-establishing the pear at a safe site, and two mature bushes were removed to the safety of a nursery. The remaining bushes may well succumb to industrial development, despite the help and cooperation of the local authorities and conservation groups. At present one tree flourishes in the grounds of a factory, whose enlightened owners guard it with care.

# · *WILD PEAR* ·

**_Pyrus pyraster_ Burgsdorff (_P.communis_ auct.)**

*Fruit*
The fruit of wild pear is extremely variable in shape, from moderately tapered, as in the cultivated forms, to almost spherical. They tend to be small, 13-40mm in diameter, borne on short pedicels 5-12mm long, which are sharply demarcated from the base of the fruit. The apex is indented and carries the five shrivelled, pointed remnants of the persistent, woolly calyx lobes. The colour of the pears is greenish yellow, even when they are fully ripe in October, and the skin is rough with numerous small lenticels. The pears are very hard in texture, the flesh full of silicaceous granules and virtually tasteless. The fruits of pear trees which have reverted from the cultivated state, although frequently misshapen, are usually longer and larger, smoother skinned and borne on longer 30mm pedicels, so that the fruits hang elegantly, and not jammed together in clusters. The seeds of wild pear are 8 × 4mm, brown, with a pointed base.

Wild pear does not usually bear heavy crops of fruit, but in the summer of 1983 one tree growing in an ancient hedge in Sussex with wild service-tree (*Sorbus torminalis*), spindle-tree (*Euonymus europaeus*) and midland hawthorn (*Crataegus laevigata*) fruited so heavily that the branches split through under the weight of pears.

*Leaf and Flower*
Wild pear is a deciduous tree growing to 20m, with a rough fissured bark and branches sometimes spiny. The young twigs are reddish brown. The long-stalked leaves are ovate with rounded bases and finely serrated edges, slightly downy when young. The flowers open from mid-April to late May, five to nine in a loose corymb, all on long stalks, 20-30mm in diameter with five slightly crinkled, white petals and purple anthers. The calyx is five-lobed, constricted below and slightly downy, the withered calyx persisting at the apex of the ripe fruit.

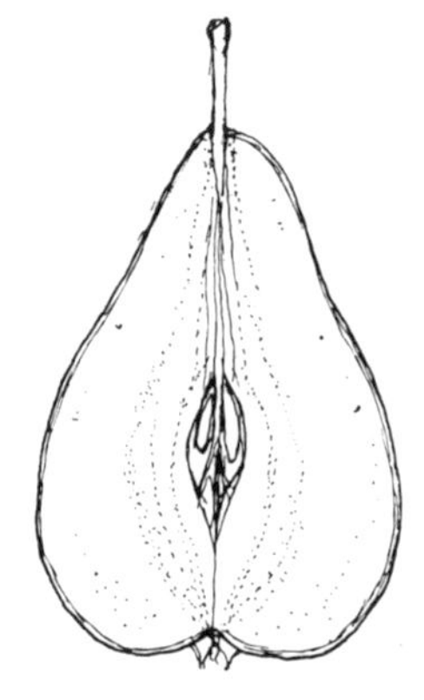

**Cultivated pear**

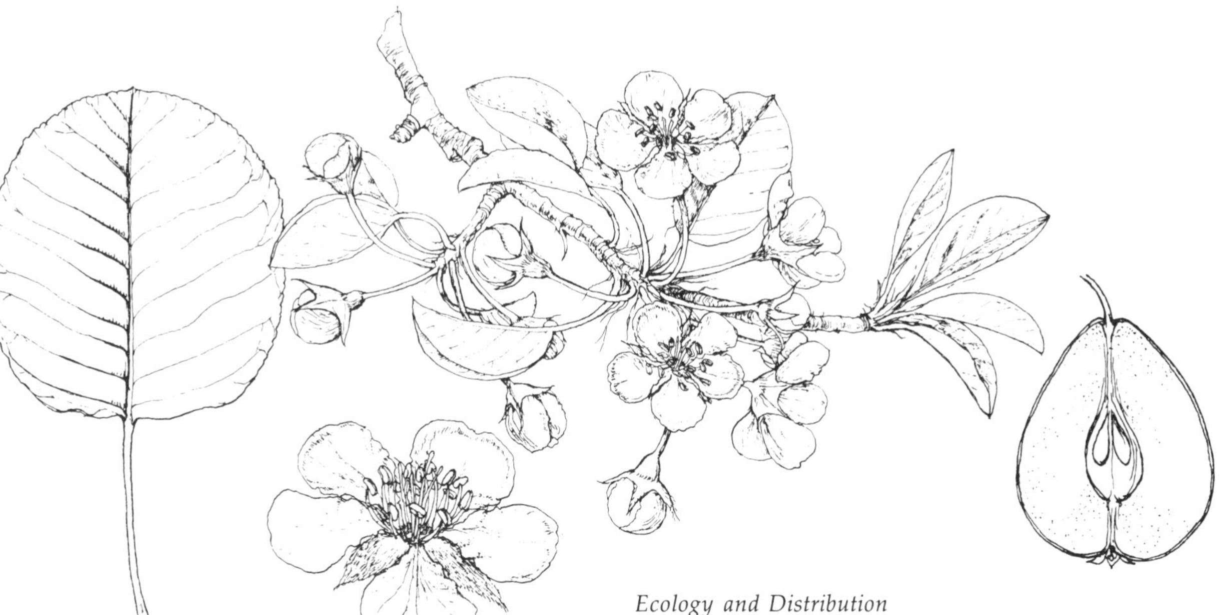

**Wild pear**

*Ecology and Distribution*
Wild pear is most frequent as a relic of cultivation, but is wild in woods and hedgerows in the south and west of England. It is absent from Scotland and very rare in Ireland, recorded from Cork, Tipperary and Antrim. In many areas it has been grubbed out, since the fruit was scarcely worth eating, and as a result it is hard to find, even in the south of England. There are still some very fine wild pear trees in Herefordshire, where they are often free-standing in fields, not growing in hedges. It was newly discovered in Alderney in 1984, and grows throughout south, west and central Europe.

*History and Uses*
The cultivated forms of pear were first developed in France and Belgium about two hundred years ago. Some authorities maintain that all 'wild pears' in Britain are relics of cultivation, but recent archaeological work has disclosed carbonized fruits in a Mesolithic midden on Téviec Island off the coast of Brittany. It is therefore quite likely that either *Pyrus pyraster* or *P.cordata* is truly native in the south of England.

Wild pears seldom ripen enough to be enjoyed raw – the local name in Sussex 'Stone Pear' speaks for itself. Culpeper wrote: 'All the sweet and luscious sorts . . . do help to remove the belly down wards, more or less . . . The said Pears boiled with a little honey, help much the oppressed stomach'.

# · CRAB APPLE ·

**_Malus sylvestris_ Miller**

*Fruit*
The fruit of the crab apple is a pome, and it is not always easy to differentiate it from the fruit of the numerous cultivated apples which escape and establish themselves wherever man casts his apple cores away.

Wild crab apples are small, 30-40mm across, sub-globose, with both ends depressed, the apex bearing the persistent calyx remnants. The fruits are hard and yellow-green until fully ripe from September to December, when many will be streaked red or turn completely scarlet with a waxy, shiny skin. When ripe they are juicy, although rather tart, with brown pips 8 × 4mm, faintly striated and pointed at the base. Other apples are never so neatly rounded and dimpled at both ends.

*Leaf and Flower*
The crab apple is a small deciduous tree, rarely reaching 10m, with a dense round crown and rough scaly bark. The branches are smooth, brown and frequently spiny. The long-stalked leaves are pointed ovate, finely serrate and smooth on both surfaces when mature. Crab apple flowers from mid-May to late June, bearing simple umbels of four to seven flowers, 25-40mm across, with five white petals suffused with pink and yellow anthers. A crab apple in full bloom is a very lovely tree, and many varieties have been developed for gardens. The calyx is smooth and hairless on the outside, although woolly on the inner surface, and persists on the apex of the ripe apple. Many of the descendants of escaped cultivated apples have leaves which are tomentose beneath, and calyces and flower stalks which are also tomentose all over.

*Ecology and Distribution*
Crab apples are common and widespread in hedges and woodland throughout the British Isles except the far north. Fruit-bearing varies greatly from year

to year, depending upon the warmth in May and early June, and the activity of pollinating bees. In northern counties this is reflected in less prolific crops, and in 1954 J.W.Heslop Harrison noted that *M.sylvestris* flowered well in most seasons in the Team Valley in Co Durham, but rarely fruited.

*History and Uses*
The name derives from the Norse 'skrab', meaning small and rough (tree). Crab apples are very tart, but make excellent jelly and cider, and are best gathered after the first frost. The fermented, concentrated juice was commonly called Verjuice, and mediaeval cooks used it as we would use lemon juice. It contains malic and citric acids, with sugars, and is a useful intestinal astringent, the fruit pulp being used to treat diarrhoea. Crab apples were also an important ingredient of the Wassail Bowl, which consisted of ale, honey and spices, to which were added crab apples roasted with a little butter. This treatment made them puff up and split open, when they were added to the heated, spiced ale. This concoction is also known as 'lamb's wool'.

Five species of birds have been recorded eating crab apples, especially during December and January, for they are essentially a hard weather food.

**Crab apple**

Blackbirds are the most frequent eaters, followed by fieldfares.

*Toxicology*
The leaves of crab apple contain the cyanogenic glycosides prunasin and prulaurasin, while the seeds contain amygdalin. Hydrolosis of these within the stomach releases hydrocyanic acid. Cases of poisoning are rare, but in September 1985 in Somerset one goat died and another was poisoned; each ate about 500g of crab apple leaves and fruit.
**Symptoms** The live goat was found in a collapsed state, breathing rapidly and with greyish mucous membranes. A sample of venous blood was as bright red as arterial blood. (Cyanide poisoning prevents tissue uptake of oxygen, and results in venous blood becoming as fully oxygenated as arterial blood.)
**Treatment** A mixture of animal charcoal and bismuth subnitrate in water was given by mouth, and sodium nitrite followed by sodium thio sulphate injected intravenously. Further oral doses were administered, and the goat made an uneventful recovery.

# · ROWAN ·

***Sorbus aucuparia* L.**

? POISONOUS

The fruits of all the members of the genus *Sorbus* are pomes; hence they have a structure like a tiny apple.

*Fruit*

The fruits of the rowan, borne in dense flat clusters, are a glorious orange-red. Trees with yellow fruits have been recorded from Bedburn, Co Durham and Bunclody in Co Wexford. Each fruit is round with a flattened top, in the centre of which the scars of the calyx attachment form a five-rayed star, each ray ending in a little dimple. The fruits are 6-9mm in diameter, and if cut across show a division into three cells, each of which can contain two seeds, 3.5mm long, pointed oval, pale brown and shiny. Usually only two or three seeds will mature. The skin of the berry is tough and shiny with a very few inconspicuous lenticels, while the flesh is rather mealy, with a sweet apple scent and a very acid taste. The fruits ripen from mid-September onwards, and can remain on the tree until December, unless stripped by birds.

*Leaf and Flower*

The rowan is a slender, smooth-barked tree 15-20m high, with silvery grey bark and pinnate leaves 100-200mm long having five to seven pairs of lanceolate leaflets, with acutely toothed margins. The terminal leaflet is never larger than the laterals. The leaflets are hairless above, whitish grey and downy below when they are young. Rowan comes into flower from late May to late June, bearing large corymbs of creamy white blossoms with a foetid scent. Each flower is 6-10mm across, with three or four styles and numerous golden stamens. The pedicel lengths vary, so that the inflorescence is almost flat-topped.

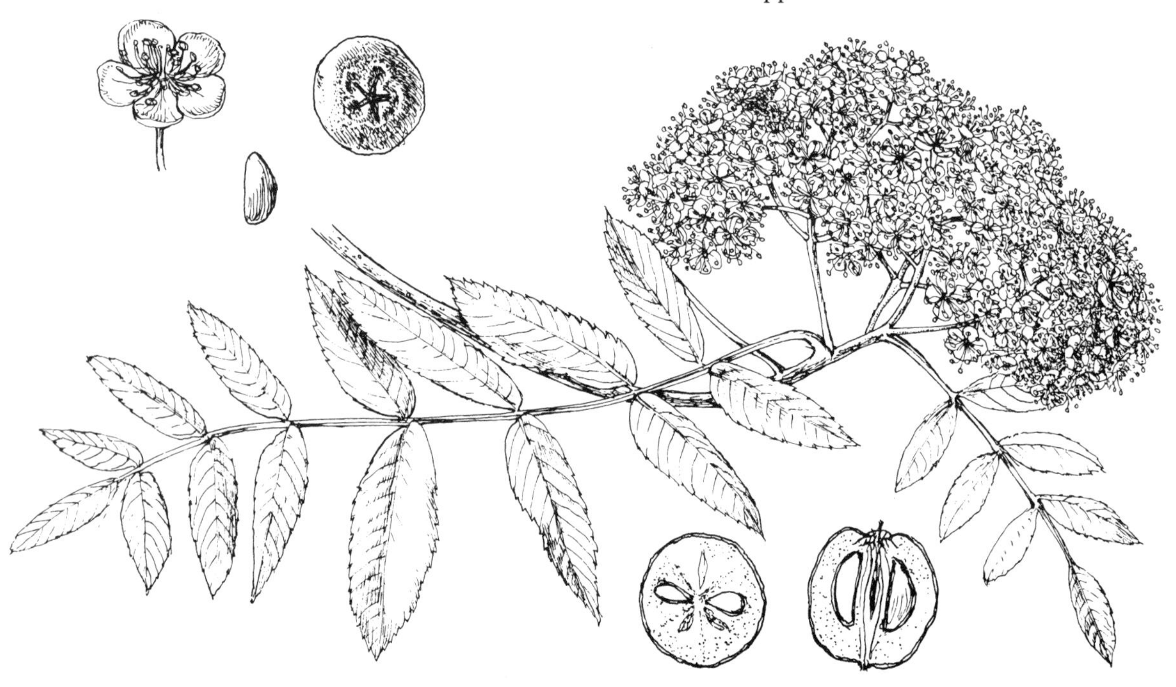

**Rowan**

*Ecology and Distribution*
Rowan is widespread throughout the British Isles, in woodlands, heaths and rocky mountain areas of lighter, non-calcareous soils. It is particularly common in the north and west at heights up to 950m, and is part of the ancient Scottish pinewood flora with birch, aspen, juniper and holly. It is absent from the coasts and chalk hills of southern England and the lowlands inland from the Wash. It grows throughout Europe, except for southern areas bordering the Mediterranean.

*History and Uses*
The name Rowan derives from the Gaelic *rudha-an*, 'the red one'; it is also known as Mountain Ash on account of the leaf shape, and as Roddon, Quicken and Witchen. The last name comes from its alleged properties in keeping away witches and evil spirits. In times past a rowan was often planted near a croft door for this purpose, or a bunch of rowan would be hung in the door of a cottage or byre, to protect the inhabitants or their animals from evil. Rowan berries are best picked in October. They are rather unpleasant eaten raw, and to prepare them for use some people soak them in a weak solution of vinegar for eight to twelve hours, or boil them for a short time and discard the water. Rowan can be used to make a pleasant-tasting jelly, or the fresh juice can be added to gin, imparting a flavour like Angostura bitters. Rowan berries can also be used to make a light red wine, and also a dye. They contain malic and parasorbic acids, sorbitol, vitamin C and sugars. The juice has been used to treat constipation and rheumatic pains. The berries frequently attract birds, of which twelve species have been recorded eating them, particularly blackbirds, mistle thrushes, redwings and song thrushes. Starlings will also eat them, especially in a hard winter.

*Toxicology*
The fruits are reported to contain a fixed oil and also a cyanogenic glycoside. There is some evidence that they can cause poisoning in children, producing symptoms of excitement, convulsions and respiratory distress. The process of boiling involved in making jelly or jam could well destroy these poisonous properties.

# · WILD SERVICE-TREE ·

***Sorbus torminalis* (L.) Crantz**

*Fruit*
The pomes of wild service-tree grow on angular pedicels in loose corymbs, like the rowan, and are easily identified by their olive brown colour when ripe. They are ovoid, 12-18mm long, slightly longer than broad, and covered with numerous orange-brown lenticels which give the skin a rough texture. The shrivelled calyx remnants persist at the apex of the fruit. In texture the fruits are hard, with rather scant dry flesh, and precious little flavour. They usually contain a single seed which when ripe resembles a large brown apple pip, 8 × 3mm, with a pronounced keel along one side. The fruits are ripe in October and November.

*Leaf and Flower*
Wild service-tree is a deciduous tree, usually of slender growth up to 13m – though large, free-standing, trees can reach 25m with a girth exceeding 4m – and rough fissured bark. The leaves are shaped much like the leaves of sycamore (*Acer pseudoplatanus*), with deeply cut, pointed lobes, green and hairless on both surfaces. In autumn they turn a spectacular flame or gold colour, rendering this rather unobtrusive tree much easier to find. Wild service-tree flowers from late May to mid-June, with corymbs of creamy coloured flowers 10-15mm across, much like the flowers of rowan. The calyx is woolly, with pointed triangular teeth, spreading and finally shrivelling as the fruit ripens.

*Ecology and Distribution*
Wild service-tree is thinly scattered in woods on clay and limestone throughout England and Wales, being most frequent in parts of the Midlands and south east England. It grows in the Weald, in Essex, in undisturbed primary woodland in Epping Forest where it is occasionally pollarded, in the New

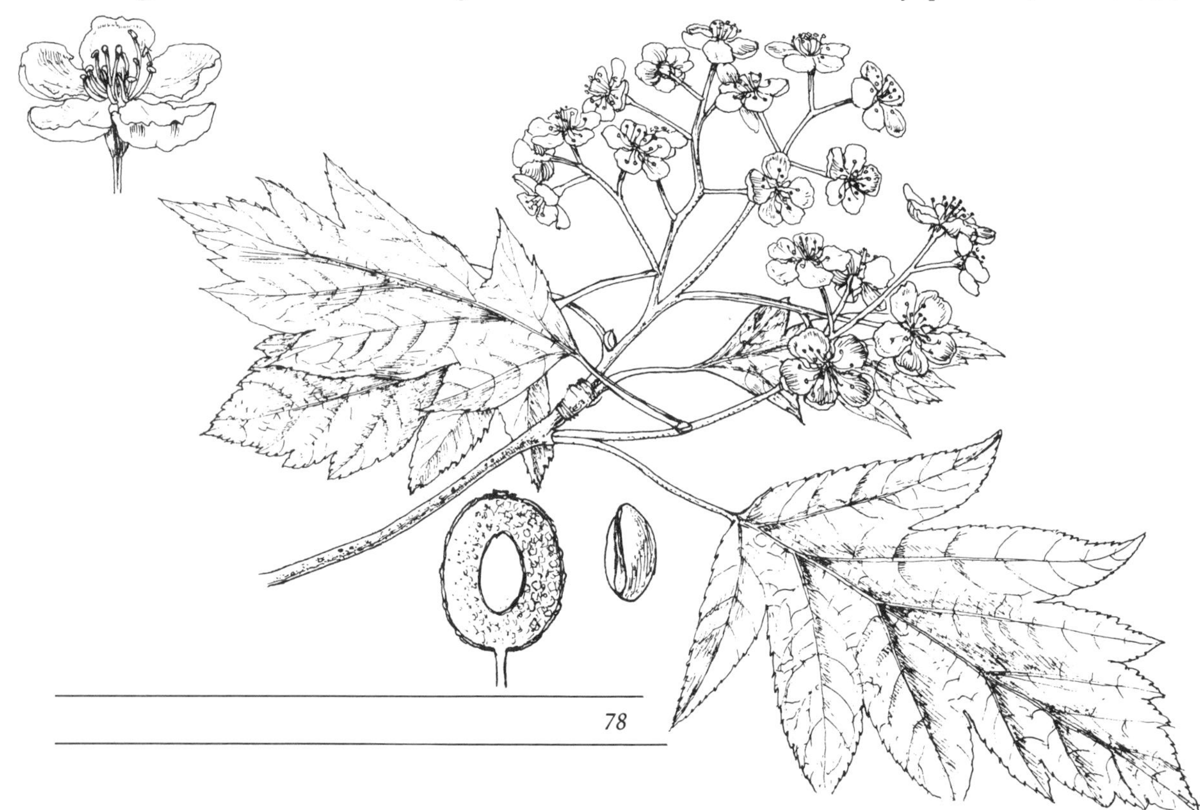

Forest, Bernwood in Buckinghamshire, three newly discovered woodland sites in Bedfordshire, Needwood in Staffordshire, and the Delamere Forest in Cheshire. It is relatively abundant in the Wye Valley both in Herefordshire and Gwent, where it grows on cliffs above the river, also in Pembrokeshire, but there is none now in north west Wales. Only one plant could be found in a recent survey in the Tamar Valley in the West Country. It grows as far north as the Humber on the east coast, and on the west coast north to Cumbria. It has a wide distribution in south, west and central Europe east to Denmark.

A survey of the trees in Epping Forest has shown that regeneration is effected principally by the production of suckers from long runner roots, extending up to 110 m from the parent tree. These are only produced by old trees, possibly over two hundred years old, and because of the curious manner in which they grow, the plants originating from these runners curve towards the parent tree. This produces a leaning trunk, so that the tree appears to be heeling over. Production of new plants from seed is erratic, and most satellite trees will prove to have originated as suckers. Seed may take two seasons to germinate, and various methods of stratification over winter, or subjection to varying periods of freezing alternating with warmth, have all resulted in depressingly poor rates of germination. Many fallen seeds are destroyed by boring insects.

*History and Uses*

The name service-tree may derive from the Latin for beer, *cerevisia*, for the Romans used the fruit of the related True service-tree (*Sorbus domestica*) to flavour it. Wild service-tree was widespread in the lowland woods of England and Wales until late mediaeval times, and the fruits were used in jams, preserves and liqueurs. As late as the 19th Century the fruits were marketed as 'Chequers', a name commonly used in the Weald of Kent. Gent (1681) noted: 'The Service-Tree, the berries are cold and binding, it is good to stay bleedings of Wounds, Lasks and Fluxes of Blood'. Other names were Surries (Lincolnshire), Lizzaries (Gloucestershire and Worcestershire), Serve or Sherve (Kent and Sussex) and Sharves (Essex). When fully ripe the cuticle darkens, and the fruit becomes soft and pulpy, with a pleasant prune-like flavour.

*True Service-tree (Sorbus domestica)*

This is not a native species, but an uncommon introduction. It looks like a large rowan, with rough bark and larger pinnate leaves with leaflets grey-green below, softly downy with whitish hairs. The newly opening buds are large, bright green and tacky, the white flowers 12 mm across opening in early May. The fruits are ripe in October and November, and are large, 25 mm long, ovoid, green flushed with brownish red, and marked with many small dots and lenticels. They have a mild taste. In Worcestershire they went by the name of 'Whitty Pear'. A famous tree first recorded in 1678 grew in the Wyre Forest near Bewdley until it was burned out, although it was replaced in 1916. One of the largest was grown by Lord Mountnorris in 1800-1820 at Arley Castle in Worcestershire, and by 1961 it had reached a height of 20 m, with a girth of 3 m. Self-sown plants have occurred rarely on calcareous soils.

**Wild service-tree**

# · The Whitebeams ·

About twenty species of *Sorbus* occur in the British Isles, and of these three species, the rowan (*Sorbus aucuparia*), common whitebeam (*S.aria*) and wild service-tree (*S.torminalis*), are sexual diploids or tetraploids, producing seed by normal fertilization and also forming hybrids within the genus. Many of the other species are polyploid and at least partially apomictic, producing seed without fertilization. When pollen is deposited on the stigma of the apomictic species the pollen tube does not grow to reach and fertilize the ovule, but the presence of the pollen grains stimulates seed formation. Thus hybrids can only arise between sexually reproducing members of the genus, or between a sexually reproducing and an apomictic species where the latter supplies the male parent – the pollen.

Many of the whitebeams recorded in the British Isles are extremely local in distribution, in some cases restricted to a single stream valley or cliffside. The whole distribution picture has also been confused by the widespread introduction of whitebeams foreign to the British Isles, for planting in parks, gardens and beside roads, and these have subsequently hybridized with native whitebeams.

The leaf shape is a vital criterion in determining the species of a whitebeam. Always look at the leaves on mature branches, or on the spurs of mature branches, never those on the end shoots. Even with these precautions, they can prove a tricky genus. In the Wye Valley in Herefordshire the common whitebeam (*S.aria*) shows a bewildering variety of leaf forms, from small oval to suborbicular, some with shallow lobes and some without lobes.

Using leaf-shape as a guide, the whitebeams have here been assigned to three aggregates, those resembling *S.aria*, those like *S.intermedia* and those like *S.latifolia*. Within each group details of individual species are given where these will help to make identification easier.

# · COMMON WHITEBEAM ·

**_Sorbus aria_ (L.) Crantz**

*Fruit*

The fruits of the whitebeams are pomes, like those of rowan. The fruit is bright scarlet and shiny, with many small lenticels scattered over the surface, giving it a slightly rough texture. The ovoid fruits are long (8-15mm), rather than broad, in an uneven corymb, on yellowish 10mm-long pedicels, which may still bear little patches of the thick white felt which clothes the immature stems. The white felt of hairs also covers the remnants of the calyx at the apex of the fruit. Beneath the leathery skin the flesh has a mealy texture, sweet in taste but rather insipid. The fruit is divided into three cells, each of which may contain two seeds like flat-sided brown apple pips 5 × 3mm, but in most fruits only two of these mature. They ripen in September.

*Leaf and Flower*

The common whitebeam is a small, spreading, deciduous tree with a dense crown, 5-10m high, very occasionally growing to 25m; or it may grow as a shrub. The young branches are downy. The leaves are broad ovate, toothed and slightly lobed, broadest at or below the middle, and densely covered with smooth white felt below. There are nine to thirteen pairs of veins which are very

***Sorbus aria***

prominent on the underside of the leaf. When the young leaves moves in the breeze the contrast between the white underside and green upper surface makes the whitebeam a very attractive tree even before flowers or fruit appear. The white, faintly scented flowers 10-16 mm in diameter are borne in large corymbs, with stiff, deeply cupped white petals and creamy pink anthers. The calyx and pedicels are densely covered with white hairs. Flowering lasts from late May to mid-June.

*Ecology and Distribution*

The common whitebeam is not infrequent on the chalk downs of southern England, as one of the components of chalk scrub, and in the Chiltern beechwoods where it grows with yew and wild cherry. It is also common on the limestones of Gloucestershire, Hereford and Gwent, and elsewhere in Wales, Ireland and Scotland it has a scattered distribution on calcareous soils. The best time of year to spot whitebeam is at the end of April, when the newly opening pale green, white-backed leaves are easy to identify. Common whitebeam grows in Europe from Ireland to Spain in the west, and in the east as far as the Carpathians.

*History and Uses*

Whitebeam fruits are edible and can be used in a similar manner to the berries of rowan. It is advisable to leave them on the tree until touched by frost, when they 'blet' like medlars, and are said to have more flavour.

*Sorbus rupicola* (Syme) Hedl.

Usually a shrub 2 m tall. Leaves dark green above, thickly white felted below, one and a half to two times longer than broad, tapering uniformly from

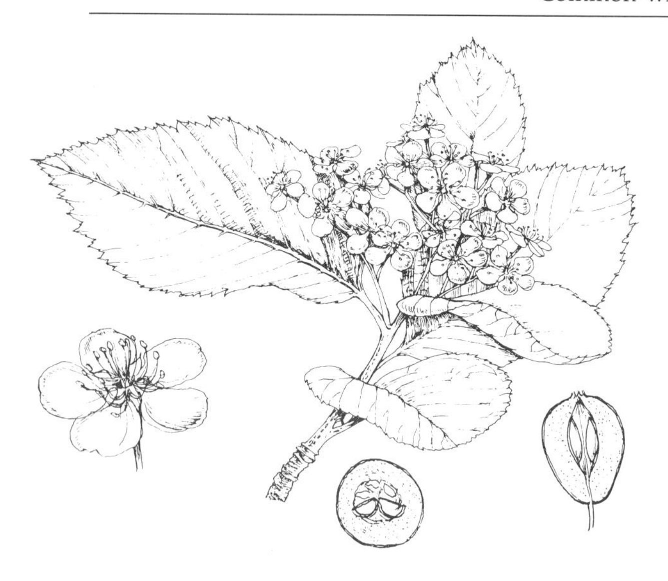

mid-leaf to pedicel. Flowers 14mm. Fruits 12-15mm broader than long, carmine with numerous lenticels mainly around base of fruit. Limestone and basic rocks to 460m. Very local from Staffs. and Derby north, South Devon, North Wales, Highlands, North and West Ireland.

*S.porrigentiformis* E.F.Warburg
Leaves rounded apex, tapering below with outward pointing teeth making angle of 90° with leaf axis. Dark crimson, sub-globose fruits with few large lenticels. Torquay, south Somerset and Mendips to South Wales, with one recent record Caernarvonshire.

*S.hibernica* E.F.Warburg
Elegant small tree with ascending branches and large obovate leaves 80-100mm, broadest above middle with rounded apex. Mature leaves dull green above, white below. Immature fruits white downy, mature fruits pinkish scarlet with numerous pale brown lenticels. Endemic Central Ireland on limestone, Burren Co Clare, north to Sligo, east to Armagh and Meath, south as far as Waterford.

*S.wilmottiana* E.F.Warburg
Avon Gorge, Gloucestershire. Fruits carmine red to scarlet, down persisting on apex, 12.5 × 11mm broader than tall. Few lenticels. Pyrenes 5.5 × 3mm with slight curve to tip.

*S.eminens*
Avon Gorge. Large obovate leaves and large fruit 20mm across.

*S.vexans*
Mendips, Somerset near to the coast. Similar to *S.rupicola*.

*S.leptophylla*
Brecon, Wales

*S.lancastriensis* E.F.Warburg
Similar in form to *S.rupicola*, characteristic of limestone rocks and cliffs, mainly in West Lancashire between Cunswick Scar and Warton Crag.

***Sorbus wilmottiana***

# · SWEDISH WHITEBEAM ·

**_Sorbus intermedia_ (Ehrhart) Persoon**

*Fruit*
The fruit of Swedish whitebeam is distinctly longer (11-15mm) than broad, scarlet in colour, with very few or no lenticels on the skin of the fruit. The fruiting heads are not broad and are usually composed of six to ten fruits. The dry calyx remnants at the apex are large and erect. The light brown seeds are 5 × 3mm, ellipsoid, with a markedly curved tip. Fruit is ripe in October.

*Leaf and Flower*
Swedish whitebeam is a smallish tree 5-10m high with a broad crown, although some venerable trees near Malham in Yorkshire are very broad and about 15m high. The leaves are elliptical, broadest about the middle, with rounded, forward-pointing lobes giving a distinctly pinnatifid appearance. The upper leaf surface is dark green, and the contrasting under surface smoothly felted with yellowish hairs, marked by eight to ten pairs of prominent veins. The white flowers appear from late May to mid-June, 8-10mm across, in broad corymbs.

*Ecology and Distribution*
Swedish whitebeam is an introduced species, widely planted in gardens and parks, and by roadsides. The attractive young foliage and succession of white flowers and red fruits make a welcome addition to any garden on calcareous or basic soil. It is most common in south east England, growing less frequently on limestone rocks in the west of England, and may be found introduced in all areas. In Europe it is widespread in the southern part of Fennoscandia around the Baltic.

*Sorbus anglica* Hedl.
Leaves broadly ovate, lobed quarter-way to midrib, with greyish woolly undersides. Fruit sub-globose with few lenticels. On limestone rocks in the Avon Gorge, Gloucestershire, the Mendips in Somerset, and widespread in Wales.

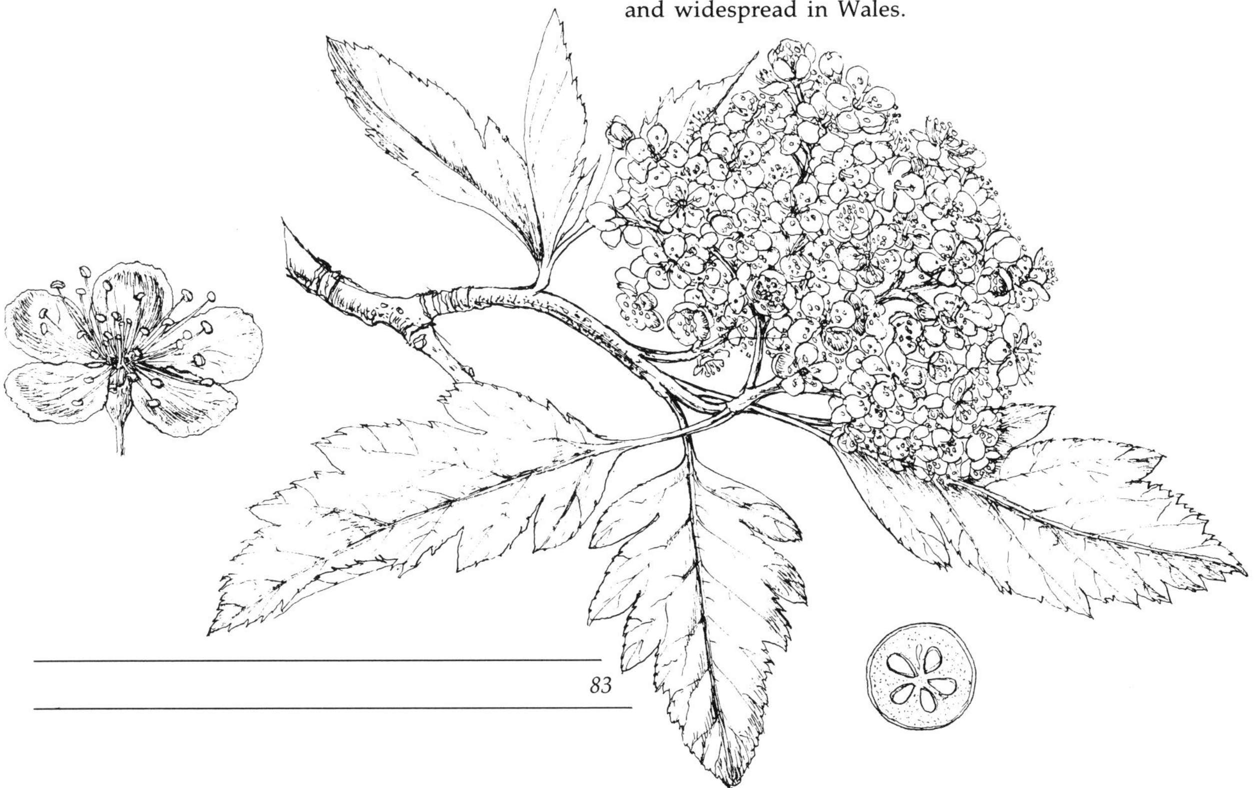

*S.minima* (A.Ley)Hedl.
Shrub 3m tall. Leaves twice as long as broad, shallowly lobed, with an even thin grey tomentose undersurface. Flowers 8mm across. Fruits small 6-8mm, sub-globose, scarlet with a few lenticels. On limestone, Craig y Cilau, Brecon.

*S.leyana* Wilmott
Leaves wider than *S.minima*, lobed at base half-way to midrib. Flowers and fruit larger. On limestone, Cwm Taf Fawr, Brecon.

*S.arranensis* Hedl.
Shrubby, leaves lobed at base half-way to midrib. Fruits ovoid and small. Stream gorges in north Arran. This and the next species are unique among whitebeams as they do not grow on base rich soils or limestones, and may have arisen as a hybrid between *S.aucuparia* and *S.rupicola*, although the latter parent no longer exists on Arran.

*S.pseudofennica* E.F.Warburg
Very similar to *S.arranensis* but some leaves show a free basal leaflet, demonstrating the kinship with *S.aucuparia*. Stream gorges in north Arran.

**Swedish whitebeam**

# · BROAD-LEAVED WHITEBEAM ·

**_Sorbus latifolia_ (Lam.) Persoon**

*Fruit*
The fruits of broad-leaved whitebeam reflect the close affinity of this group with wild service-tree (*S.torminalis*). They are usually fairly broad, rounder than those of wild service-tree, 9-15mm long, brownish or orange-brown, shiny and covered with numerous large lenticels. The calyx remnants are prominent at the apex of the fruit. They are ripe in October. The pyrenes are smooth and brown, 5.5 × 4mm.

*Leaf and Flower*
Broad-leaved whitebeam is a medium-sized tree with broad ovate leaves, rounded at the base, with seven to nine pairs of veins and broad triangular lobes with pointed ends, or with very prominent straight teeth at the end of the main veins, again showing its close relationship to *S.torminalis*. The underside of the leaf is greyish and tomentose. The creamy flowers, 20mm across, open in May and June.

*Ecology and Distribution*
It is a rather local plant found in woods and hedges in south-west England on calcareous soils and among limestone rocks. It is sometimes planted and naturalized. In Europe it grows in Portugal, Spain, west France and south west Germany.

**Sorbus latifolia**

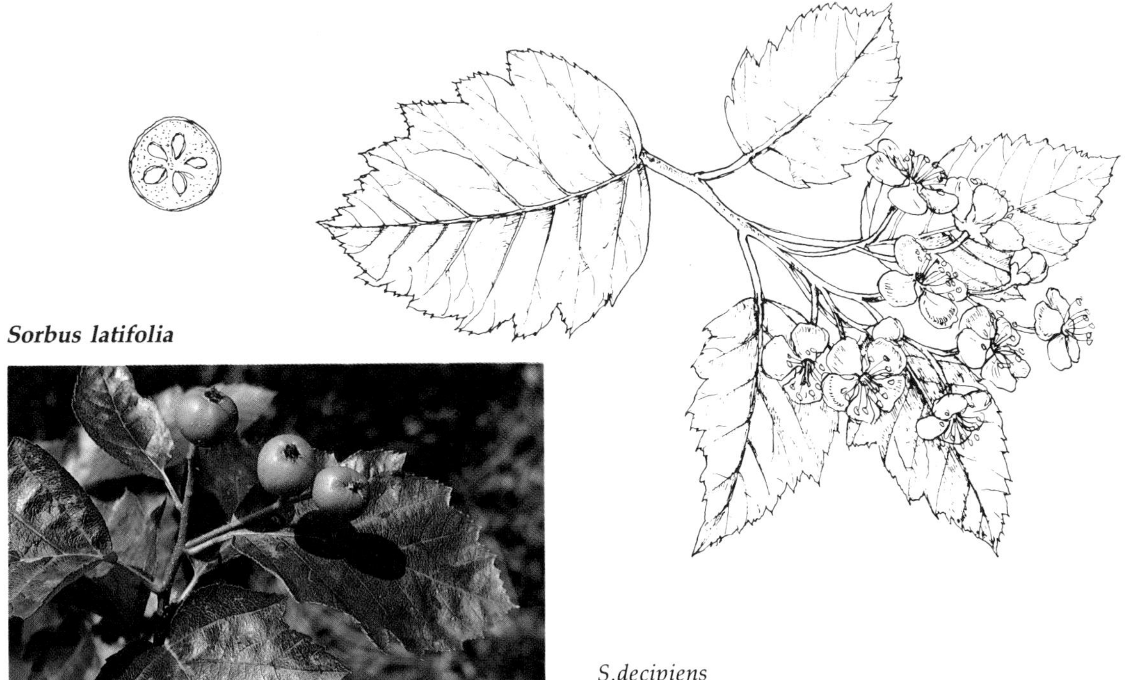

*Sorbus devoniensis* E.F.Warburg
Leaves ovate, rounded at the base, with shallow triangular lobes, grey-green below. Fruit subglobose. Grows in Devon, east Cornwall and on limestone rocks in the valleys of the Barrow and the Nore in Carlow and Kilkenny, Ireland. It has been called French Hales and Devon service-tree, and the fruit is occasionally sold and eaten in Devon.

*S.bristoliensis* Wilmott
Leaves obovate, shiny above and greyish tomentose below; teeth almost at right-angles to main leaf axis. Flowers with pink anthers. Fruit just longer than broad, 11mm long, shiny, bright orange with small lenticels all over the surface. Pyrenes 4.5 × 3.5mm. Limestone rocks in the Avon Gorge, Gloucestershire.

*S.subcuneata* Wilmott
Leaves elliptical, narrowed below, whitish on the underside. Flowers with cream anthers. Fruit subglobose, brownish orange. Limestone rocks in the Mendips in Somerset, and near the coast in north Devon.

*S.decipiens*
Leaves densely felted grey below, teeth on the lobes jagged and well developed. Fruit almost globular 12mm, dull surface, greenish-yellow, tinged with brown. Pyrenes 4.5 × 3.5mm.

**Sorbus bristoliensis**

# · *JUNEBERRY* ·

## ***Amelanchier lamarckii*** **F.Schroeder**

There has been considerable confusion on the taxonomy of *Amelanchier* in Europe, where at least three species have been naturalized since the 18th Century, all undoubtedly of American origin. There appears to be a single species present in southern England, and all the British material is referable to *Amelanchier lamarckii* F.Schroeder. Although this is of American origin, it has been overlooked in the USA as an apparent hybrid. It contains characteristics also found in *A.canadensis, A.laevis* and *A.arborea.*

*Fruit*

The fruit of juneberry is a flattish, black berry, green when first formed, turning red and finally purplish-black when ripe in September and October. The berries are 10-12mm in diameter on pedicels 10-20mm long; they are swollen just below their attachment to the fruits, which are borne in clusters near the main stem in the axils of the side shoots. The berries are shiny, with minute lenticels which give the sides of the fruits a slightly duller, rough and frosted appearance. The apex is depressed with the whitish scar of the calyx remnants. The berries are juicy, with a sweet taste.

*Leaf and Flower*

Juneberry is a deciduous shrub or small tree reaching 12m in height, with finely toothed, pointed oval, yellow-green leaves on long pedicels. There are prominent alternate veins on the underside of the leaves, whose edges tend to curl inwards. The flowers are white, with five upright, narrow petals 15-18mm across, carried sparingly in loose clusters. They appear in April, with the newly opened leaves which at this time are copper tinged, with silky white hairs underneath.

*Ecology and Distribution*

Juneberry is planted and naturalized in woods and scrub in southern England, and is particularly well established in woods in the Hurtwood area of Surrey. In 1978 it was found growing in Cumberland. It grows also in north Holland, north west Germany, Belgium, Denmark and Sweden.

*History and Uses*

It was introduced as *Amelanchier laevis* from America in 1800 under the name of 'Shad Bush'. There the fruits were eaten raw, or cooked in pies, puddings and jellies. Other names for it were Shadberry and Serviceberry.

*Amelanchier spicata* (Lam.)K.Koch, sensu Fernald non Jones

Also called *A.humilis*. A tree usually 4m tall, leaves thickly white downy when young, having finely serrate margins. Two other species may be included here – *A.stolonifera* Wieg. and *A.canadensis* (L.) Medic., the latter also known as *A.oblongifolia*. This type is found in Belgium, Austria, north west Russia and north to central Scandinavia.

*Amelanchier confusa* Hyland
Also called *A.grandiflora*. A tree to 9m, the young leaves purplish, and the flower spikes drooping. Flowers and fruits are both large and carried on long pedicels. Three other species may be included here – *A.laevis* Wieg., *A.alnifolia* Nutt. and *A.arborea* (Michx.)Fern. This type is found in central and southern Sweden.

*Amelanchier ovalis* Medic.
Also called *A.vulgaris* and *A.rotundifolia*. This is a shrub growing to 3m, with black bark and rounded leaves with coarsely serrated margins. It is supposed to occur in the south of France.

**Juneberry**

# · WALL COTONEASTER ·

***Cotoneaster horizontalis* Decaisne**

*Fruit*
The fruits of wall cotoneaster are borne almost sessile in dense clusters on the upper surface of the herring-bone-shaped branches. The berries are scarlet, rather dumpy, about 7mm in diameter, with the apex puckered in by the five-pointed scar of the calyx remnants. The skin of the berry is shiny and tough while the flesh has a mealy texture like that of hawthorn berries, and a sweetish taste. The berries contain two or three pale brown seeds, 4mm long, which are pyrenes, each rather flat and sharp edged with a keel on one side and a roughened area on half the other side.
The berries are ripe in September and may stay on the bushes well into February of the following year.

*Leaf and Flower*
Wall cotoneaster is a deciduous, low-spreading bush 0.5m high, with stiff branches spreading horizontally in a herring-bone pattern. The young twigs are covered with brownish woolly hairs, especially on the undersides. The leaves open early in the year and are numerous and densely crowded on the stem. They are rounded oval, with a sharply pointed apex, glossy above and smooth below. Tough in texture, they turn orange and then red in autumn, before falling, leaving the red berries still on the bushes. Flowers are produced singly or in pairs in May and June, with small erect pink petals and an abundant nectar supply which attracts hoards of bees.

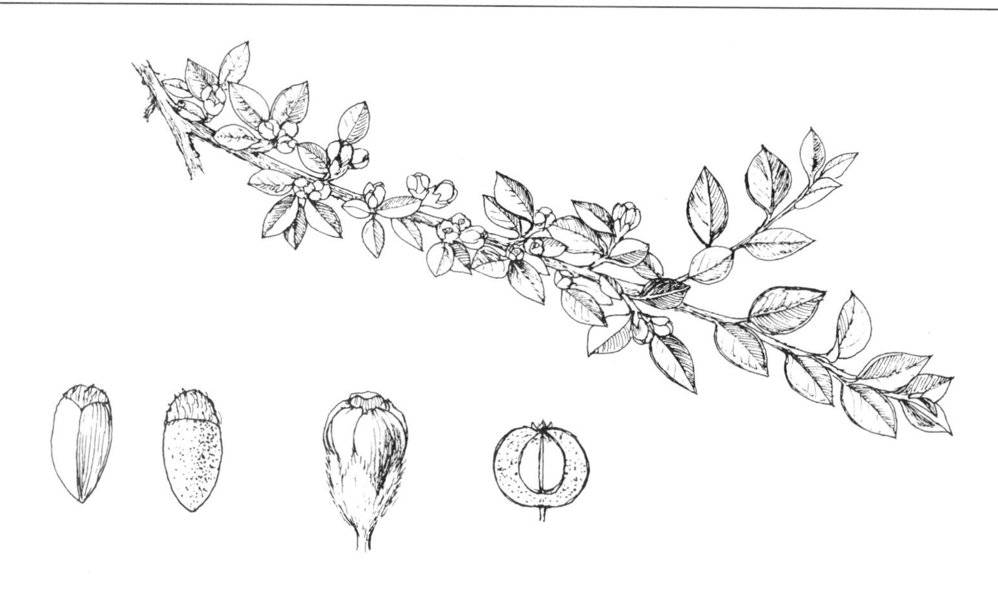

**Wall cotoneaster**

*Ecology and Distribution*
It is frequently grown in gardens, less frequently naturalized from bird-sown seed. It is more common in south England on chalk and limestone and calcareous sand, but has been recorded north in Cumberland and as far as Sutherland; on limestone in North and South Wales; and in Killarney in south west Ireland. It is also naturalized in Austria.

*History and Uses*
Wall cotoneaster was introduced from west China during the late 19th Century, and has since spread widely in the British Isles. The fruit is of no value to man but is avidly devoured by birds; foremost among the eleven species recorded eating it between August and March are blackbirds, song thrushes, bullfinches, greenfinches, house sparrows and waxwings. In years of waxwing irruptions into Britain a good way of finding them is to frequent areas where there is plenty of cotoneaster.

*Toxicology*
There is no substantiated reference to poisoning in man with any of the cotoneaster species established in the British Isles.

# · HIMALAYAN COTONEASTER ·

***Cotoneaster simonsii*** **Baker**

*Fruit*

The fruits of Himalayan cotoneaster are carried in loose, drooping clusters of four to ten on pedicels 2-3 mm long. The berries are 8 mm long, 7 mm wide, rather four-sided, and orange red in colour. The pedicel, berry base and apex all bear long felted hairs, while the equator of the berry is shiny. The apex is flattened and crowned with the shrivelled calyx remnants. Each fruit contains two or three shiny brown pyrenes, irregular and somewhat flattened, 5 × 2.5 mm. Half of one side of the seed is yellow and rough. The fruits ripen in September and remain on the bushes until the turn of the year or later.

*Leaf and Flower*

Himalayan cotoneaster grows to 4 m, with erect stems, and with twigs covered in brown down especially when young. The leaves are usually deciduous, pointed oval, about 30 mm long, with a dark green slightly downy upper surface and a lower surface felted with white hairs. Unlike wall cotoneaster, the leaves remain when the fruit is ripe, and some may not fall even with the onset of winter. The flowers open early in groups of two to four from late February, and continue until June, with erect, slightly incurved, white petals marked with red.

**Himalayan cotoneaster**

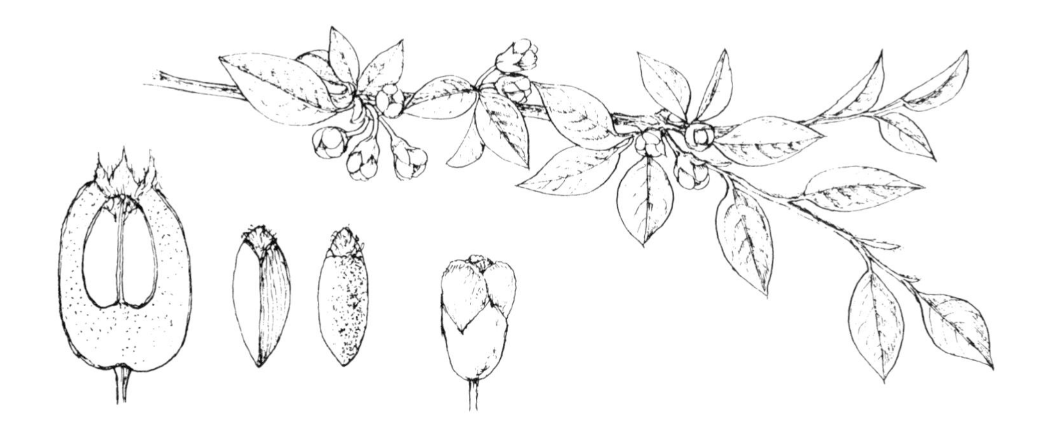

*Ecology and Distribution*
Himalayan cotoneaster is naturalized in coppices and waste places and is not infrequently bird-sown, having a partiality for scrub areas on chalk and limestone. It is now widespread in southern England and Wales and has been recorded as far north as Sutherland and west to the western seaboard of Ireland. As is the case with so many alien cotoneasters it is easily established and persistent. It is naturalized in France and Norway.

*History and Uses*
Himalayan cotoneaster was introduced from the Khasi Hills south of Shillong in Assam by a Mr Baker in about 1865. Its provenance was confirmed when it was refound for certain in Assam in 1886 by C.B. Clarke. During the autumn and winter from August to March twelve species of bird have been recorded eating the berries, especially blackbirds, song thrushes, bullfinches, green finches, house sparrows and waxwings. There is even an odd record of skylarks eating them in January.

# · WILD COTONEASTER ·

***Cotoneaster integerrimus* Medicus**

*Fruit*
The fruits of wild cotoneaster are carried in small, drooping clusters of two to four, crimson in colour, somewhat pear-shaped, with an indented apex crowned with prominent calyx remnants. The skin of the berries is rather dull, like that of small crimson haws. There are two to four pyrenes, 3 × 4.5mm, three-angled, yellowish and finely striate. The fruits are ripe from mid-August to the end of the year.

*Leaf and Flower*
Wild cotoneaster may be a small, spreading bush up to 1m high, but is usually compact and prostrate, with thick, slow-growing stems which are downy when young but become smooth and grey with age. The leaves are deciduous, entire and rounded-ovate, 10-40mm long, glabrous and deep green on the upper surface, grey tomentose beneath. In young leaves a fringe of these grey hairs is visible around the leaf edge. The flowers open from May to June, 4-6mm in diameter, with five pink, erect petals which are often incurved. The flowers are borne in drooping clusters of two to four, and the calyx is hairless. They are pollinated by wasps, but at high altitude may be self-pollinated.

*Ecology and Distribution*
Wild cotoneaster is more strictly calcicole than the other cotoneasters and is extremely rare in the British Isles, where it is our only native cotoneaster. It has been known for many years from the carboniferous limestone cliffs of Great Orme's Head near Llandudno, Gwynedd. Elsewhere in Europe it is not uncommon in dry, stony calcareous areas in the Alps, Pyrenees, in Scandinavia, central and eastern Europe. Its range extends to north Asia, where it is not uncommon in Zanskar and Kashmir at altitudes of 2,600-4,000m. The photograph shown here was taken at 4,000m on limestone cliffs near Sheshnag in eastern Kashmir. Another species recorded there as *C.integerrima* is now called *C.falconeri*.

*History and Uses*
The finding of wild cotoneaster on Great Orme's Head in 1783 was attributed to J.W. Griffith, but this is unlikely, as he appears to have been born in that year. It was rediscovered in 1821 by William Wilson of Warrington, who described it as plentiful on ledges above Llandudno and elsewhere on the Orme. It suffered severely at the hands of collectors and visitors, so that by 1908 W. Gardner could only find nine plants. By 1924 G.C. Druce could only find five. In 1970 constructive efforts were made to

**Wild cotoneaster**

save the species, and from the surviving four plants M. Morris collected seeds and cuttings for propagation. Later, with the help of T. Parry of the local cliff rescue team, seven individual plants were reintroduced to sites which were not too exposed either to grazing animals or the elements. By autumn 1982 two of these had survived, the others probably succumbing to desiccation, and in the summer of that year four mature three year-old plants were introduced, and seed was sown in five localities. By 1986 two of the 1970 batch of transplants and the four transplants of 1982 were still alive and healthy, and a further two mature plants had been introduced. There was no evidence of successful germination of planted seed, but a further two mature original plants had been discovered in a previously unrecorded site.

The berries of wild cotoneaster have been used as a treatment for diarrhoea, and there has been a suggestion that they were grown at Great Orme for medicinal purposes.

*Toxicology*

There is an unsubstantiated report that the berries have caused poisoning in children, with symptoms of excitement, convulsions and respiratory distress, possibly due to a cyanogenic glycoside.

# · *SMALL-LEAVED COTONEASTER* ·

***Cotoneaster microphyllus*** **Wallich ex Lindley**

*Fruit*

The fruits of small-leaved cotoneaster are relatively large, 9mm long, globose or broadly pear-shaped, carried singly or in pairs on short 2mm pedicels. Production of fruit is usually rather sparse, and they tend to be obscured by the stems and foliage. The berries are bright scarlet with a tough skin which is not shiny but is minutely pitted. The apex of the fruit bears large, prominent calyx remnants. There are usually two pyrenes 5.5mm long in each fruit, all of them having a shiny flat side and a curved side partly rough and pitted. Fruits are ripe from September to early November.

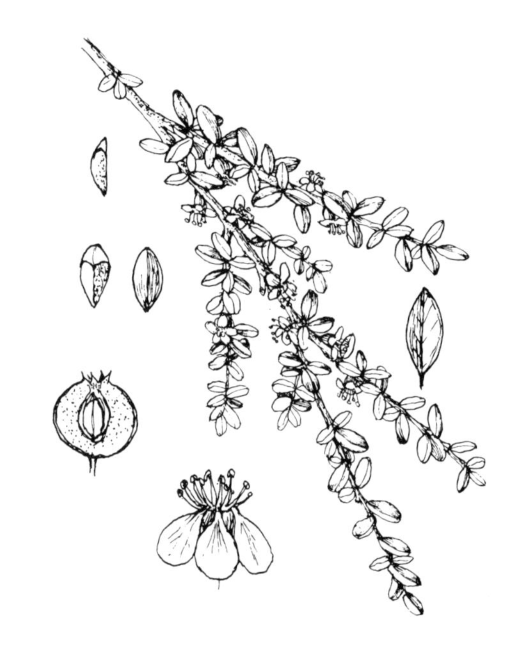

*Leaf and Flower*

Small-leaved cotoneaster is a small bush up to 1m high, with straggly, arching stems, which are downy when young and often grow flat on the ground, partly hidden in the grass. The shrubs are usually evergreen, with small ovate leaves 5-10mm long, broadest near the tip, dark green and shiny above, and felted with long appressed hairs beneath. Some of the leaves may have an apical notch. The flowers open from mid-May to mid-June, usually solitary, with white petals which are spreading or reflexed. The white stamens are long and prominent, tipped with purple anthers.

*Ecology and Distribution*
Small-leaved cotoneaster is grown in gardens and is naturalized chiefly on limestone cliffs and chalk downs. It has a scattered distribution throughout the British Isles as far north as Lochinver in Sutherland and Lewis in the Outer Hebrides, but is most common in southern England, Wales and on the western limestones in Ireland, where it has been seen growing on Abbots Hill, Co Clare.

*History and Uses*
Small-leaved cotoneaster is an alien, introduced from the Himalaya; it is still widespread there and in south west China. Following its introduction into England in 1824, by 1905 it had spread to Glamorgan and by 1920 it had appeared in Sussex, Gloucestershire and Brecon. The last sixty years have seen it establish itself from one end of the British Isles to the other, though it is seldom abundant.

**Small-leaved cotoneaster**

# · MEDLAR ·

**_Mespilus germanica_ L.**

*Fruit*
Medlars are easy to recognize by their unusual shape and texture. They are carried singly on short stalks, each fruit 20-50mm across, shaped like a truncated pear. The top is wide open in a saucer shape rimmed with the five-pointed, hairy calyx lobes 10-15mm long, and exposing the five carpels. Medlars are honey brown in colour, densely covered in a fuzzy felt of hairs. Each of the five carpels contains a wedge-shaped stone 12 × 3mm, with woody, wrinkled walls. The flesh is hard and rather woody, and although in the Mediterranean the fruits ripen on the tree, in Britain it is usual to wait until they are half rotten, or 'bletted', usually in late October about the time of the first frosts.

*Leaf and Flower*
The medlar is a small, rarely thorny, deciduous tree, 2-6m high, with a smooth bark. The leaves are lanceolate, dark green and downy especially on the underside, and have very finely serrated margins. In autumn they turn yellow. Flowers are produced from May to late June, 30-40mm across, solitary on short pubescent pedicels. The petals are white and crinkled and at the centre are five styles surrounded by the stamens with brick red anthers. The lobes of the calyx are hairy, long and pointed, persisting around the open apex of the ripening fruit.

*Ecology and Distribution*
Medlar is naturalized in hedges, mainly in the south of England where it is very local and decreasing, usually growing as isolated trees which are easily overlooked. Some trees are of considerable age, such as that first noted in Surrey in the 19th Century by John Stuart Mill at Redstone Hill, Redhill; and one at Godstone Mill Pond. Another tree first recorded at Clevedon undercliff in Somerset in

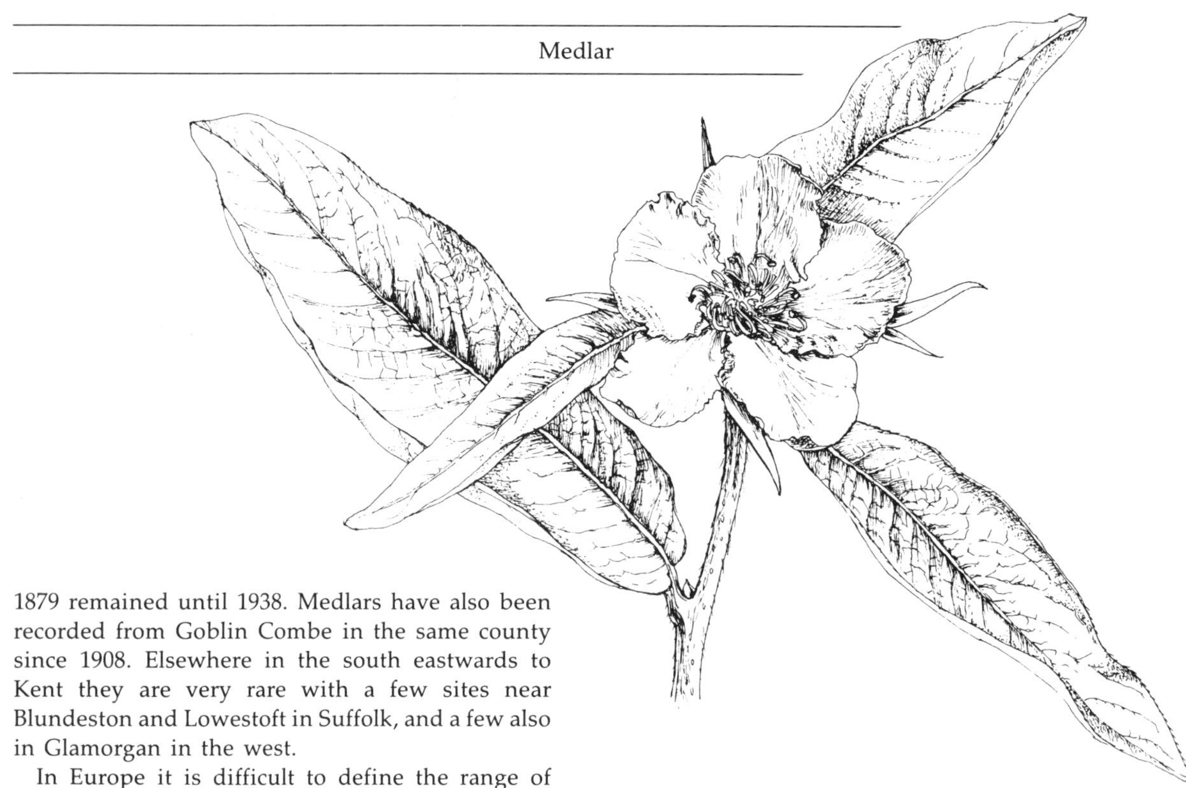

1879 remained until 1938. Medlars have also been recorded from Goblin Combe in the same county since 1908. Elsewhere in the south eastwards to Kent they are very rare with a few sites near Blundeston and Lowestoft in Suffolk, and a few also in Glamorgan in the west.

In Europe it is difficult to define the range of native plants, since it has been cultivated there and in the Near East for at least two thousand years. Although widely distributed, it is likely to be native only in the south east Balkans, Caucasus, Crimea, Asia Minor and north Iran.

*History and Uses*

Medlars in the British Isles rarely ripen and need to become 'bletted' in order to be edible; even then they are an acquired taste. They can be eaten raw, baked or for making jelly. Culpeper wrote: 'They are very powerful to stay fluxes of blood or humours in men and women . . . The fruit if eaten by pregnant women stays their longing for unusual meats'. He also mentioned their use in poultices and in the treatment of renal calculi. This use is also mentioned by Gent (1681): 'Medlars are cold, dry and binding, the Leaves are of the same Nature, they are good to stop all Fluxes of Blood inwardly given, and the dried Leaves beaten to powder and strowed upon bleeding Wounds, stayeth the bleeding of them, and healeth them up quickly; the stones of them made into powder, and given in white-Wine, wherein Parsley Roots hath layn, infusing all Night, doth break the Stone in the Kidneys, and help to expel it'.

**Medlar**

# · MIDLAND HAWTHORN ·

**_Crataegus laevigata_ (Poiret) DC**

*Fruit*
The haws of midland hawthorn are bright scarlet and shiny, slightly dumpier and flatter topped than those of hawthorn (*C.monogyna*), particularly in those which contain more than one stone, when the haw will be more broad than long. The haws, 8-10mm, are never so numerous as those of hawthorn and are often carried erect on moderately long pedicels in groups of five to ten. The five pointed calyx lobes are prominent on the top of the haw, and from the depression between them the stylar remnants protrude like the antennae of an insect. There are one to three styles, usually two, and they may not be easily detected.There will be one stone to each style, 6 × 5mm, and they can be so closely applied to each other that they appear as one. In determining the species of a hawthorn it is necessary to examine plenty of fruits before making a decision. The haws contain a slightly mealy, buffish-pink flesh, which has a mild, sweet taste. The stones are helpful in identification, as they are distinct from those of hawthorn, having a broad cap, and a single deep sulcus on either side. The fruits are ripe from early September onwards, remaining on the bushes well into December.

*Leaf and Flower*
Midland hawthorn is an erect shrub or small tree, 2-10m high, and it is less thorny than hawthorn. The leaves are slightly leathery, shiny on the upper surface, with broad, forward-pointing lobes (usually three), which are divided less than two-thirds the distance to the midrib. The leaf veins meet at the midrib nearly at right angles to the leaf axis. The leaves are paler than those of hawthorn and fall earlier in the autumn, leaving fruits on bare twigs. The flowers open in May to mid-June, usually eight to ten days earlier than those of hawthorn, in few-flowered cymes of five to ten. The flowers are larger, 15-20mm in diameter, with pinkish-white petals which are often incurved, and pinkish-purple anthers.

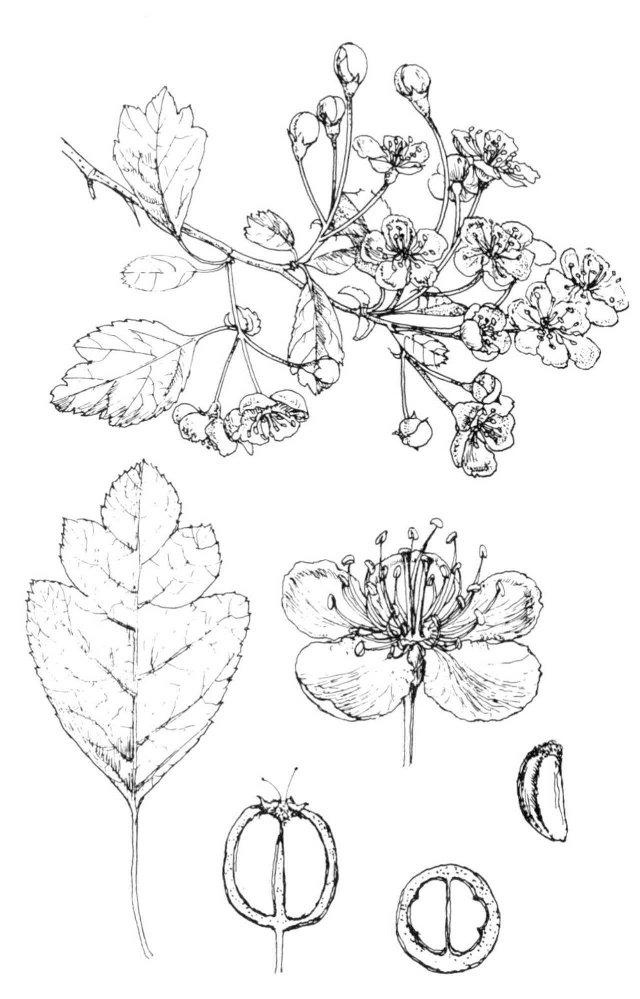

*Ecology and Distribution*
Midland hawthorn is a woodland plant and favours heavy clay soils. It is widely distributed and not uncommon in southern England and the Midlands, extending in the south west as far as Evercreech in Somerset, northwards in scattered colonies mainly in South Wales, to Cheshire, Westmorland and Angus in Scotland, where it is likely to be an introduction. In Ireland there is one tree in a wood

**Midland hawthorn**

at Kilgobbin, Dublin, where it may have been introduced; it has also been recorded near Macroom in Co Cork growing in a hedgerow with *C.monogyna*. In 1885 a specimen of the hybrid *Crataegus laevigata* x *monogyna* was found, and it was concluded that *C.laevigata* had been lost through introgression and the destruction of Irish woodlands in the 16th and 17th Centuries. Midland hawthorn is widespread in north west and central Europe.

Hybrids with *C.monogyna* occur where clearings enter woodland and where hedging with *C.monogyna* is used. Under these circumstances the normal ecological separation of the two species breaks down, and fertile hybrids result. In much of south east England most populations contain hybrids, and the situation has been reached where populations on similar soils bear a strong relationship to each other. Pure stands of *C.monogyna* still exist on the chalk hills, and pure *C.laevigata* occurs on the clay.

*History and Uses*

*C.monogyna* has been used extensively for hedging in the last three centuries. Before that *C.laevigata* grew in hedges created by forest clearance in Saxon times. Hedges on clay soils, with pure *C.laevigata* populations, can well be over one thousand years old. Unlike hawthorn, midland hawthorn soon gives off a stink of putrid flesh when plucked, due to the presence of trimethylamine, and this may be the reason for fertility beliefs becoming powerfully associated with it. It flowers earlier than hawthorn, which would be unreliable for May Day celebrations, and it is possible that midland hawthorn was the original May Flower.

The fruits contain trimethylamine, ursolic and aleanolic acids, with purine derivatives. They act as cardiac sedatives, dilating blood vessels, and leading to a fall in blood pressure. The dried flowers, fruit and leaves of both hawthorn species contain flavonid glycosides with supposedly cardiotonic properties. Under the name 'Crataegus' they were used especially in Italy, France, Holland, Germany and Japan. The fruits can be used in jam, but have little flavour.

# HAWTHORN

**_Crataegus monogyna_ Jacquin**

*Fruit*

The haws of hawthorn are scarlet and shiny when first ripe in early August, turning duller and more crimson the longer they remain on the bushes. They are 6-12mm long, long rather than broad, crowned with five wide, blunt calyx teeth from the centre of which protrudes the single style which gives the plant its scientific name. The fruits are more shiny than those of midland hawthorn, with orange-tinged, mealy flesh and a single yellowish stone 8 × 5mm. The stones can be differentiated from those of *C.laevigata* by the narrow band and cap at the apex, and the presence of two sulci on each side. Plants with yellow berries have been recorded rarely.

**Hawthorn**

*Leaf and Flower*

Hawthorn is an erect shrub or small tree 2-10m high, thorny and deciduous. The leaves are usually dull on the upper surface, with three to seven lobes which are bluntly pointed and divided more than halfway to the midrib. The leaf veins tend to radiate from the leaf base. The flowers are carried in large, flat, strongly scented corymbs of ten to twenty. The five petals are creamy white, sometimes pink, and the numerous stamens carry pinkish-purple anthers. The scent of hawthorn is strong and heavy, particularly in warm, moist weather, and can be unpleasantly cloying.

Hawthorn has two distinct flowering periods, the main season from late May to mid-June and another from mid-November to January, with fruits being formed only from the summer flowers. Southern

European forms may flower earlier in summer than the English form, even when moved to the north, but they do not flower twice.

*Ecology and Distribution*

Hawthorn is a common plant throughout the British Isles, in open woods, hedges and thickets, growing up to 550m. It is less common in the far north of Scotland, absent from Orkney, and uncommon in Shetland. It is the dominant scrub plant of chalk downs and limestone grassland. During the time of the Enclosures, hawthorn was grown in vast quantities to provide good, quick-growing, stock-proof hedges, a use reflected in the names 'Whitethorn' and 'Quick'. It is common throughout Europe, except on the far northern and eastern boundaries.

*History and Uses*

The use of hawthorn as hedging material has been mentioned already, the fortunes of several of our oldest horticultural families having been made in the production of the millions of 'quicksets' planted. The name derives from the Old English word 'Hawe', which referred to a space enclosed by a hedge. In times past, ancient thorn trees were used as public meeting places, such as the Hethel Old Thorn in Norfolk. They were often held to be sacred, and linked to Puck or the Fairy Folk, especially in Eire.

One of the most fascinating legends concerns the Holy Thorn of Glastonbury in Somerset, which is supposed to have sprung from the staff of Joseph of Arimathea. The story tells how Joseph reached Glastonbury in AD63, and when he thrust his staff (itself an offshoot of Christ's Crown of Thorns) into the ground, it promptly burst into flower. The first reference is in an anonymous poem of 1520, which starts: 'Here begynneth the Lyfe of Joseph of Armathia'. Apparently there were three thorn trees on Wearyall Hill, south west of Glastonbury, which 'do burge and bere grene leaves at Christmas'. There is no mention of winter flowering. The first Christmas flowering was recorded in 1535. In the reign of James I, cuttings were taken and dispersed as a garden curiosity, and in the ensuing Civil War this saved the strain from extinction, as the surviving 'original' tree was destroyed by the Roundheads. The Glastonbury Thorn is undoubtedly a form of *C.monogyna*.

Hawthorn flowers can be used to flavour liqueur or to make wine, and the berries can be steeped in brandy to make a liqueur. The haws make a moderately tasty jelly if supplemented with crab apples. In mediaeval times a pudding was made of 'Flowrys of Hawthorn', and the berries and seeds were used medicinally. Culpeper wrote: 'The seeds in the berries, beaten to powder, being drank in wine are held singularly good against the stone and are good for the dropsy'. Gent in 1681 also recommended them: 'The Hawthorn-Berries are very binding, therefore are good to stop a Lask, the Berries dried and drunk in white-Wine is very good against the Stone and Dropsey, and seed bruised after it is cleared from the Down and drunk, is good for the tormenting pains in the Belly'. The haws contain trimethylamine, ursolic and aleanolic acids, with purine derivatives, and an extract can be used as a cardiac tonic and diuretic.

Young leaf shoots can be eaten in salads, and buds can also be used to make a pudding, hence one of hawthorn's odder names, 'Chucky Cheese'.

Haws are an important source of food for wood mice and other small mammals, as well as supplying food for birds in winter. Twenty-three species have been recorded eating them, chiefly blackbirds; also redwings, fieldfares, mistlethrushes, song thrushes and waxwings.

*Toxicology*

There have been some unsubstantiated reports of poisoning in children, causing excitement, convulsions and respiratory distress. The presence of a cyanogenic glycoside has been suggested.

# · CHERRY PLUM ·

***Prunus cerasifera* Ehrhart**

*Fruit*

The fruit of cherry plum (which is also known as Myrobalan) is a drupe. It hangs singly on a long, 15mm pedicel like a cherry, but is much larger, 20-30mm in diameter, slightly longer than broad and well marked with a shallow groove down one side like a plum. Cherry plums are usually cherry-red in colour and shiny, although plants with yellow fruits have been recorded. They have juicy flesh and a pleasant flavour, and are ripe in late July. The stone is relatively small, rounded and smooth. Fruit is not produced every year.

*Leaf and Flower*

Cherry plum is a deciduous bush or small tree growing to 8m. It has slightly drooping branches with spiny ends, the young twigs greenish, glossy, and usually thornless. The leaves are pale green, glossy above, lanceolate with serrate margins. Some plants are hairy on the lower side of the midrib and leaf stalk. The flowers open in mid-March with the leaves (it is the earliest plum to flower) and are usually borne singly; they are white and 15-22mm in diameter.

*Ecology and Distribution*

Cherry plum has been introduced and widely planted in hedges and shrubberies, mainly in south-east and central southern England. It has also been recorded in east Scotland from Berwick and East Lothian. In Wales it is an established alien in several counties, and more recently has been found between Cynwyd and Corwen in Merioneth, and

Kelsterton in Flintshire. It has recently been recorded (1983) in Co Armagh in Ireland. It is native to the Balkans and Crimea, eastward into Asia, and has been introduced into Denmark, Austria, Germany, France and Italy.

*History and Uses*
The date of this tree's introduction into the British Isles is not known. It is used not only for the fruits, which can be eaten raw or cooked, but also as a stock for grafting.

*Toxicology*
In common with all the members of the genus *Prunus*, the seeds within the stones contain the cyanogenic glycoside amygdalin.

**Cherry plum**

# · *BLACKTHORN* ·

***Prunus spinosa*** **L.**

*Fruit*
The fruit of blackthorn is the sloe, a small blue-black plum 9-15mm long. It is a drupe, with a shiny skin which often has an attractive blue bloom and greenish flesh with a highly astringent taste. As the pedicels are usually short – less than 5mm – the masses of fruit are closely packed on the stem, but there is a small-fruited form, forma *pauperis*, with fruits usually solitary, black and shiny, carried on longer pedicels 9-10mm. The stone within the sloe is smooth and rounded. Sloes are ripe from late September to November.

*Leaf and Flower*
Blackthorn is a bush or small tree growing to 4m and frequently suckering to form dense thickets. The bushes are deciduous, with dark, rigid branches armed with vicious spines at the tip, making the trimming and clearing of blackthorn scrub a daunting task. The thorns are a frequent cause of penetrating wounds in animals, particularly when they have walked through an area where hedge-trimming is in progress. The thorns are so hard and sharp that they will pierce the sole of a horse's foot or cow's hoof, causing severe sepsis. Blackthorns have a just reputation in the countryside of being 'dirty' thorns. The leaves are pointed oval, broader near the tip, with serrated margins. The leaves of forma *pauperis* are half-sized, very narrow and pointed. Blackthorn is in flower from March to late May, the flowers opening well before the leaves. They are carried in axillary clusters on short, smooth pedicels and have five pure white petals, 10-15mm in diameter. A bush in full flower is a glorious mass of dazzling white blossom set against bare dark twigs – one of the first signs of spring, but the term 'blackthorn winter' is a reflection of the bitter weather that can so often return in March, just as the blackthorn flowers. On sea-side shingle, especially in south east England, blackthorn grows in a dwarf, prostrate form, with flowers, leaves and fruit identical to the common form. These prostrate bushes are prodigious bearers of fruit, so much so that in a good season the bush disappears under its

**Blackthorn**

**Form on shingle**

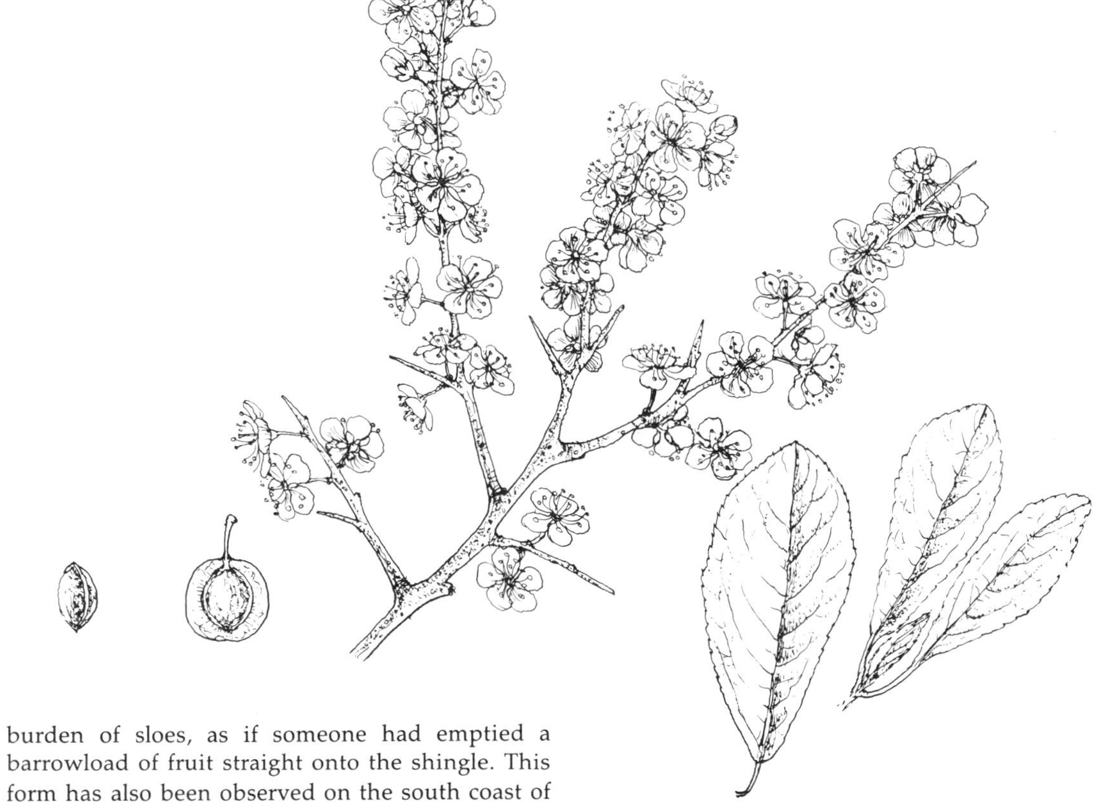

burden of sloes, as if someone had emptied a barrowload of fruit straight onto the shingle. This form has also been observed on the south coast of Mull.

*Ecology and Distribution*
Blackthorn is common and widely distributed throughout the British Isles, except for some parts of the Highlands of Scotland, Orkney and Shetland. It grows in scrub woodland and hedges, on sea cliffs and in fens, often forming impenetrable thickets. The normal form and forma *pauperis* occur side by side, while the prostrate form is restricted to shingles. Blackthorn is common throughout Europe, except for the north east and extreme north.

*History and Uses*
Sloes have been eaten and used for a very long time, but their juice is so highly astringent that it instantly dries the mouth, so that no one would eat raw sloes for pleasure. However, they make a superb jelly and are widely used for making a liqueur (sloe gin). They are best picked just after the first frost, since this renders the skin softer. To make sloe gin, prick the sloes and pack them into bottles with an equal weight of sugar, filling the bottles to halfway. Top up the bottles with gin and seal tightly, storing for at least three months, during which time they should be gently shaken to ensure that the fruit and sugar mix properly. The contents should then be strained and bottled, and may be drunk at Christmas, although sloe gin will improve with keeping, if given the chance. Culpeper wrote: 'The fruit is chiefly used, and is restringent and binding, good for all kinds of fluxes and haemorrhages. It is serviceable in washes for sore mouths and gums, to fasten loose teeth . . .' Gent (1681) agreed – 'The Black-thorn or Slo-bush, all the parts of it is cooling, and binding, and drying, and good to stay Bleeding at the Mouth and Nose, stop the Lask of the Belly or Stomach, bloody Flux, and to ease the pains of the Bowels and Guts, that come by overmuch Scourings, the Leaves are good to put into Lotions, to wash a sore Mouth or Throat with, wherein are Sores and Kernels, and to stay the deflections of Rheums to the Eyes or other parts, and to cool the heat of them'.

The juice of unripe sloes, boiled to reduce it to a thick consistency, was used in the 17-18th Century as an intestinal astringent and called German Acacia. The juice of ripe sloes can be used as an indelible marking ink. Sloes are occasionally eaten by blackbirds.

# · WILD PLUM ·

***Prunus domestica* L.**

*Fruit*
Wild plum is widespread as a naturalized species, usually as a relic of cultivation. The colour of the fruit therefore varies greatly, from yellow-green to purple, depending on the form from which it is derived. The skin may have a bloom on the surface. Size is also variable, the plums ranging from 20-80mm in length, usually ovoid or egg-shaped. They are consistently more lop-sided than the closely related bullace, with a marked, one-sided sulcus. The flesh is usually sweet and juicy, and separates cleanly from the stone when the plum is ripe in September. The shape of the stone is helpful in distinguishing wild plum from bullace, as it is smooth and fairly flat, with a well defined sharp edge.

*Leaf and Flower*
Wild plum grows as a deciduous bush or small tree, rarely reaching 10m. The branches seldom have spines. The leaves are very variable in shape, usually broad ovate with serrated margins, dull on the upper surface, and bearing a few scattered hairs on the veins beneath and on the pedicels. The flowers open at the same time as the leaves, from mid-April to mid-May and are 15-25mm in diameter, with white petals.

*Ecology and Distribution*
Wild plum grows in woods and hedges, usually as a relic of cultivation, remaining where old cottages have long since disappeared. It is widespread in England, Wales, southern Scotland and Ireland, especially in the warmer south east, but is less

common in the west of the British Isles. In the north it becomes increasingly rare, with one old record from Creich in Sutherland, and none from the Hebrides, Orkney and Shetland. It is widely naturalized throughout Europe.

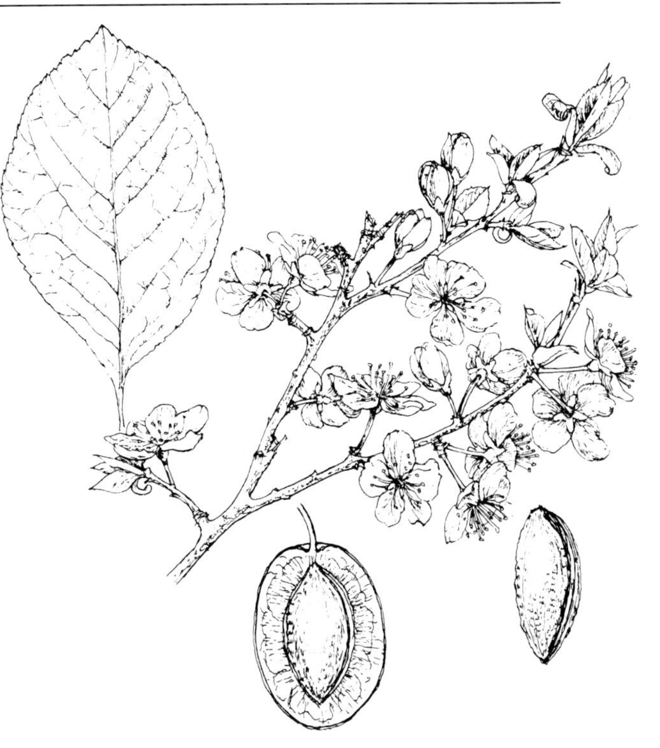

*History and Uses*
Plums were probably introduced to the British Isles from the Caucasus. They were widely cultivated as dessert fruit, cooked and eaten in pies and puddings, and used for making wine and brandy. The dried ripe fruits are called prunes, which have purgative properties and are an ingredient of senna confection.

*Toxicology*
The seed within the stone contains the cyanogenic glycoside amygdalin.

**Wild plum**

# · BULLACE ·

**_Prunus domestica_ subsp. _insititia_ (L.) C.K.Schneider**

*Fruit*

The fruit of the bullace never reaches the large size of some wild plums, being 15-25mm long, usually far more rounded, with symmetrical halves and a shallow sulcus on one side. The fruits are usually borne in small clusters on pedicels about 5mm long. The colour varies from greenish-yellow to red, and many bullaces are streaked and mottled. The flesh is usually greenish orange when ripe in September, with a fairly strong 'plum' flavour, and not very sweet. The stone is rather fat, 12 × 10mm, rough surfaced with a blunt edge; for this reason the flesh does not readily separate from the stone.

*Leaf and Flower*

The bullace is a deciduous bush or small tree, 2-6m high, with branches which are sometimes spiny. The young twigs are downy. The leaves are variable in shape, ovate with sharply serrated margins, and, like the twigs, downy on the underside when young. The white flowers, 15-30mm in diameter, appear with the leaves from early April to mid-May.

*Ecology and Distribution*

Bullaces are rather common in hedges and scrub throughout most of the British Isles, especially in the south. To the west they are widespread in north Somerset and South Wales. In the north they are less common, being rare in Angus and absent from Sutherland. In Europe bullaces are widely naturalized in woods and hedges; it is probable that, like wild plum, they originated in the Caucasus.

*History and Uses*

Bullaces have been widely used over the centuries, both as dessert fruits and cooked. One of the earliest English references is by William Palerne (1350), who wrote: 'Gete vs ... bolaces and blakeberies þat on breres growen'. The origin of the name bullace is not clear, since it occurs in many forms including the old Breton 'bolos', the Welsh

**Bullace**

'bwlas' and the Irish 'bulistair'.

The bullaces at present in cultivation have been grouped into three forms by Grieve.

*Royal Bullace* 35mm long, green mottled with red on the side nearest the sun. Green flesh separates easily from the stone.

*White Bullace* Small and round, pale yellow or white mottled with red. Juicy flesh adheres to the stone.

*Essex Bullace* Larger than white bullace. Green, turning to yellow when ripe, juicy and not so acid.

# · WILD CHERRY ·

**Prunus avium (L.)L.**

*Fruit*

Wild cherries can vary in colour from bright red to mainly yellow with red streaks. They are 9-12mm in diameter, globular and shiny, carried in drooping clusters on slender pedicels 20-50mm long. The stone is 6mm in diameter, globular and smooth, set in slightly orange-coloured flesh which may be sweet or bitter tasting but is often rather insipid. They ripen at the end of June and during July.

*Leaf and Flower*

The tree is deciduous and grows as tall as 25 metres with shining brown bark which peels horizontally into strips. The leaves are elliptical with acutely pointed tips and serrated margins. The young leaves may be copper coloured when they first open, flaccid and drooping, with a sparsely hairy underside. The mature leaves have a matt surface, and turn into beautiful shades of red and pink in autumn. On the petiole, just below the leaf blade,

are two red knobs or glands. The flowers appear from late March to mid-May, carried on long drooping pedicels in umbels of two to six without a common stalk. They are 15-30mm across, with fragile white, notched, petals and subacute sepals which soon fold back. The top of the calyx tube is constricted.

*Ecology and Distribution*
Wild cherry grows in woodland and on field margins, especially in beechwoods, favouring the glades and rides where there is more light. It is common throughout most of the British Isles, except for the north of Scotland, although it has been recorded as far north as Caithness. It is absent from the Hebrides, Orkney and Shetland, and local in Ireland. In Europe it is widespread and common except in the extreme north and east. It is rare in the Mediterranean area.

*History and Uses*
The fruits of wild cherry or gean can be eaten raw and used for making pies and tarts, although they are never as succulent as the cultivated cherries, and may be bitter-tasting. A tree in full flower is a wonderful sight, perfectly evoked by A.E. Houseman in *A Shropshire Lad*:

Loveliest of trees, the cherry now
Is hung with bloom along the bough,
And stands about the woodland ride
Wearing white for Eastertide.

The fruits are occasionally eaten by blackbirds.

*Toxicology*
The seed within the stone contains the cyanogenic glycoside amygdalin.

**Wild cherry**

# · DWARF CHERRY ·

## ***Prunus cerasus* L.**

*Fruit*

The fruits of dwarf cherry are usually carried singly on erect, stiff pedicels 25 mm long, at the base of the twigs well below the leaves, and not in groups among drooping leaves, as in wild cherry (*P.avium*). The colour of the shiny fruit varies from yellow to red as the fruit ripens in July, and may be dark crimson. The cherries are fat and juicy, 15 mm in diameter, globular with both ends deeply indented, and very tart tasting. The stone is smooth and flattened, 9 × 7 mm, with a blunt end and rough, sharp edges.

*Leaf and Flower*

Dwarf cherry is a small, deciduous, bushy shrub, usually 2-5 m tall, suckering freely, with dark green leaves, glossy above, which make the tree look sombre. The leaves are thicker and erect, shorter and rounder than the leaves of wild cherry, with irregular serrations on the margins, without the pair of red glands on the petiole. The flowers, which appear from late April to mid-May, are 18-23 mm across, on stout pedicels in small umbels, with thick white petals, scarcely notched at the edge. The calyx tube is campanulate, not constricted at the top, and the sepals are blunt and ovate.

*Ecology and Distribution*

Dwarf cherry is a plant of hedgerows and scrub, probably introduced from south west Asia, and uncommon on rather acid soils in southern England (including a sighting in Bedfordshire in 1983), Wales (especially in the south west and on Anglesey), and in Ireland. Elsewhere in Europe it is cultivated for its fruit, and is widely naturalized.

*History and Uses*
Dwarf cherry, also called Morello or sour cherry, is used commercially to make cherry liqueur. It is also used as a colouring agent, and for making cherry juice and cherry syrup.

**Dwarf cherry**

*Toxicology*
The seed within the stone contains the cyanogenic glycoside amygdalin.

# · *BIRD CHERRY* ·

***Prunus padus* L.**

POISONOUS

*Fruit*
The fruit of bird cherry is small, 5-8mm across, globular, black and shiny, resembling a black pea on a long stalk. It is quite juicy when ripe in July and August, with bright green flesh which has an unpleasant foetid odour and a foul taste which dries the mouth. The stone is relatively large, pointed oval, 6 × 4mm, with a rough, wrinkled surface. In most years very few mature cherries form from each flowering spike.

*Leaf and Flower*
Bird cherry is a deciduous bush or tree up to 15m high, with bare, brown, peeling bark which is foetid. The leaves are pointed ovate, smooth, serrate, and hairy along the underside of the midrib. The petiole bears two red glands like those of *P.avium*. In autumn the leaves turn a rich red. The white flowers, 10-15mm across, have petals and sepals with serrated edges. They bloom from mid-May to mid-June, and are unlike those of any of the other cherries in the British Isles, carried in heavily scented, spike-like, drooping racemes of ten to forty blooms.

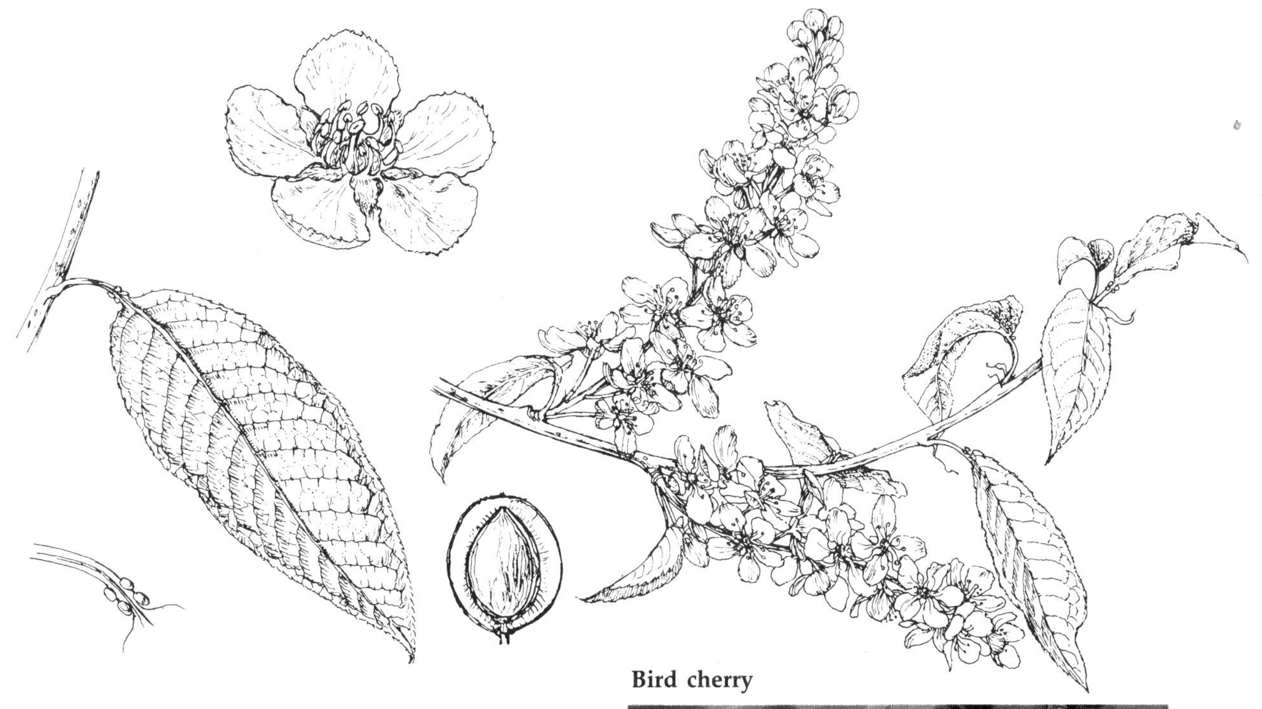

**Bird cherry**

*Ecology and Distribution*
Bird cherry grows in woods and by streams, frequently on basic or calcareous soils up to 600 m. It is common in the north of England, especially in the Dales of Derbyshire and Yorkshire, in Wales and in Scotland. J.W. Heslop-Harrison noted in 1958 that in the Team Valley in Co Durham it flowered well but rarely produced fruit. In the dry summer of 1984 many of the bushes in the Derbyshire dales showed symptoms of drought damage which, combined with a heavy infestation of moth caterpillars, resulted in an almost total absence of fruit. Apart from a local concentration of sites in east Suffolk and Norfolk, it is absent from southern England as a native plant, although this most attractive tree is widely cultivated in parks and gardens. It is less common in the north of Scotland, growing mainly in the south and east, and it is absent from the Outer Hebrides, Orkney and Shetland. It thrives best near water, and is particularly abundant by the side of Clunie Water south of Braemar, where, in contrast to the shrivelled plants in Derbyshire, it fruited heavily in August 1984. In Ireland it is rare and scattered. In Europe it is widespread, except in countries bordering the Mediterranean, in the Balkans and south east Russia.

*Toxicology*
All parts of the plant are poisonous, particularly the leaves and fruits; they contain the glycosides prunolaurasin and amygdalin. The latter is cyanogenic, and on hydrolysis breaks down to produce hydrogen cyanide and benzaldehyde, so that ingestion will result in the symptoms of cyanide poisoning.
**Symptoms** Cyanide poisoning in humans causes restlessness, staggering gait, convulsions and respiratory distress, or death without premonitory symptoms. However, the amount of cyanogen in bird cherries is so small that these classic symptoms are rarely produced. Children are unlikely to eat a significant quantity of the fruits because of their foul bitter taste.

An interesting comment on the toxicity of bird cherry was made by Yekaterina Stepanova, Deputy Director of the Tashkent Botanic Garden, USSR. This intrepid lady ate both leaves and fruits of bird cherry as an experiment to test all the plant species in their collection for possible effects on humans. Symptoms of dizziness, with marked palpitations and difficulty in breathing were rapidly produced, but fortunately subsided within half an hour without treatment.
**Treatment** Vomiting should be induced, followed by gastric lavage as a precautionary measure. Amyl nitrate is of value to stimulate the heart, and a respiratory support system should be on hand. If convulsions develop they must be controlled. NPIS have recorded one case, which was asymptomatic and the patient recovered spontaneously. No cases of animal poisoning have been recorded in Britain.

# · *HOLLY* ·

***Ilex aquifolium* L.**

POISONOUS

*Fruit*
Holly berries are globose drupes, each scarlet berry – 7-12mm long – usually divided into four locules and containing a single pendulous ovule. They are brilliantly red and shiny, densely packed in clusters of as many as sixty on a single twig, borne close to the stem on very short pedicels. The apex of each berry bears four low scars of the sepal attachments. The flesh of the fruit is orange coloured and rather slimy. Usually three or four of the ovules will mature to produce large smooth stones, each with two flat angled faces which fit neatly together. Berries with six and (rarely) eight cells have been recorded, and holly trees with yellow fruits occur occasionally. The berries ripen from mid-October and may remain on the trees until the end of February. In some years no fruit is produced.

*Leaf and Flower*
Holly is a shrub or small tree 3-15m, rarely over 20m high, evergreen, with smooth, thin grey bark and green twigs. The leaves are alternate, dark green, glossy and leathery, oval with wavy cartilagenous margins armed with large, spine pointed teeth; those at the top of the trees may be spineless. The leaves last for up to four years on the tree, rarely as long as eight years. Holly is normally dioecious, with male and female flowers on different trees, the flowers being carried on the two year-old wood in few flowered axillary cymes blooming from May to August. The flowers have four waxy white petals conjoined below, 6mm in diameter and often tinged with purple. The male flowers have four stamens and vestigeal ovaries, while the female flowers have four stigmas in the form of a disc, and vestigeal stamens. Bees are frequent visitors to the flowers, and a male tree must be near to the female tree for berries to be produced. Berry production is reduced by excessive shade.

*Ecology and Distribution*
Holly is widespread throughout most of the British Isles, in woods, hedgerows and thickets, from sea level to 550m. It is less common in central England, absent from Caithness, and rare in Orkney and Shetland. In Europe it is more common in the south and west, extending north east to north Germany

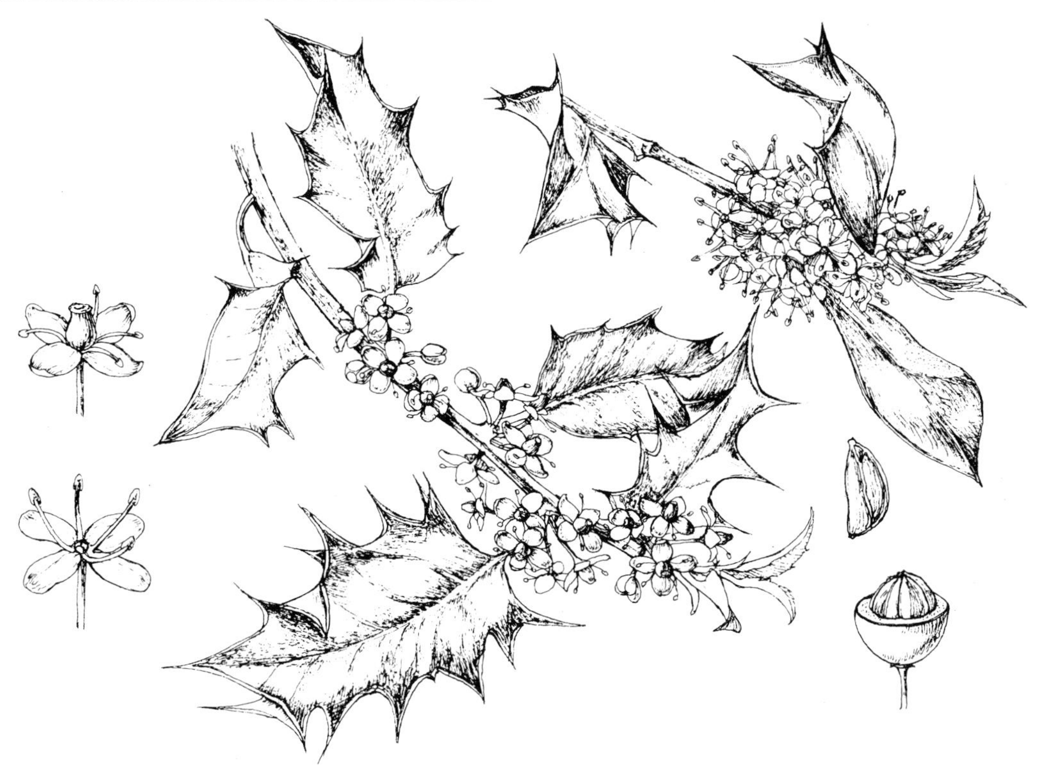

and Austria; it is now extinct in Sweden. It is usually considered to be intolerant of very wet soils and yet flourishes in situations as different as bogs in the New Forest and dry shingles at Dungeness in Kent. Holmstone Wood at Dungeness has existed as a holly wood for at least 1200 years and is mentioned in an 8th-century manuscript. The size and shape of the wood have remained unaltered since 1617. Pure stands of holly are characteristic of the New Forest, also at Stiperstones in Shropshire, and in parts of Scotland.

*History and Uses*

The ritual use of holly originated in pre-Christian times, possibly in association with the feast of Saturnalia. It was a custom among the Romans to send boughs, accompanied by other gifts, to friends. Pliny advised the planting of a holly tree near a house to protect it from witchcraft. In 1664 John Evelyn gave a recipe for birdlime made from holly. Before the 19th Century there was a wide demand for holly timber, pollarding being practised from the Middle Ages onwards and the leaves used as winter feed for cattle and sheep. Culpeper wrote: 'The berries expel wind, and therefore are held to to be profitable in the cholic. The berries have a strong faculty with them; for, if you eat a dozen of them in the morning fasting, when they are ripe and not dried, they purge the body of gross and clammy phlegm, but if you dry the berries and beat them into powder, they bind the body, stop fluxes and bloody fluxes.' Gent also approved of their use to provoke and expel wind. Deakin mentioned the manufacture of birdlime from the bark of holly.

The Holly Blue butterfly (*Celastrina argiolus*) uses holly as a feed plant, the females of the spring brood laying their eggs among the flowers. Holly berries are also an important source of food to birds, which strip the berries as soon as they ripen. At least seven species have been recorded as eating them, especially mistle thrushes, blackbirds, wood pigeons, collared doves, redwings, fieldfares and song thrushes. The birds will usually start eating at the top of the tree and strip the berries, working downwards day by day. The trees seem to vary in their season of ripening or edibility, for one will be stripped of its fruit and a neighbouring tree left completely untouched.

*Toxicology*
Although birds eat holly berries with impunity, they are poisonous to man if large quantities are ingested. The berries and leaves contain a tannin, the bitter principle ilicine, theobromine and other bitter substances. The bark contains a tannin, pectin, and the yellow pigment ilexanthin. These are irritant to the stomach, violently emetic and purgative.
**Symptoms** Holly berry poisoning in humans produces violent vomiting and purgation or diarrhoea, and may occasionally cause drowsiness. Prompt treatment should be administered to any child who has eaten more than ten berries, or an adult who has eaten more than ten to twenty berries.
**Treatment** Vomiting should be induced promptly, followed by gastric lavage and the administration of gastric sedatives. NPIS have recorded many cases, but few have resulted in serious poisoning and no deaths have been reported. No cases of poisoning involving animals have been recorded.

**Holly**

# · SPINDLE-TREE ·

***Euonymus europaeus* L.**

POISONOUS

*Fruit*
The coral-pink fruits of spindle-tree are easy to identify, each a capsule 10-15mm wide, divided into four rounded lobes. The fruits are carried in pendant clusters below the leaves on short, angled pedicels, the clusters in turn borne on paired shoots which form a right angle with the parent stem. The fruits ripen from September to November, each lobe splitting to reveal a single, rounded-oval seed 5-6mm long, covered with a bright orange membranous aril. Bushes with white fruits have been recorded in Suffolk and elsewhere. They contain the normal, bright orange-coated seeds, but when these are grown on the next generation can revert to type. Spindle-trees do not fruit consistently but in a good year, such as the summer of 1982, bushes laden with pink fruit were to be seen throughout south east England.

*Leaf and Flower*
Spindle-tree is a slender, deciduous shrub 2-6m high with smooth grey bark, the new twigs of the year being bright green and four-angled. The leaves are opposite, lanceolate and slightly toothed, turning attractive shades of pinkish-red in autumn. Occasionally plants may be found with buds and leaves in threes not in opposite pairs. Flowers appear in May and June and are small, four-petalled, greenish-white, 8-10mm across, in axillary cymes of three to ten. The four stamens alternate with the pointed petals, and a single pointed style projects from the centre of the flower. The petals soon fall, leaving the four rounded calyx lobes markedly reflexed, each flower then resembling a tiny battle mace.The flowers secrete nectar and attract numerous bees and flies.

*Ecology and Distribution*
Spindle-tree is widespread in woods, thickets and hedges, especially on chalk and limestone in the south of England. It is absent or very rare in the Midlands and the Fens, but occurs more commonly in some non-calcareous areas such as the Weald clay and over parts of the Old Red Sandstone in Hereford and south Shropshire. It is a local plant in Scotland north to the Solway and the Forth, but is virtually absent from the Highlands. In Ireland it is widespread, rare in Co Dublin but common in Co Wicklow. S.B. Evans records spindle-tree growing near Milford Haven in Pembrokeshire on an oyster midden, an isolated calcareous spot in an otherwise non-calcareous soil.

It is widespread in Europe except in some Mediterranean islands, and in the far north. The black-bean aphis *Aphis fabae* overwinters on spindle-tree.

**Spindle-tree**

*History and Uses*
In the past, the dense, close-textured mature wood was used for the manufacture of spindles. Nowadays spindle-tree is most likely to be grown for its ornamental value. Birds will occasionally eat the fruits, blackbirds being the most frequently recorded species.

*Toxicology*
The bark, roots and berries are all poisonous, containing a complex mixture of toxic substances; these include the heteroside euonoside, the cardiac glycoside euonymin, glyceriltriacetate, the pigments physaline and phyllovodine, three rhamnose sugars and the aglycone digitoxigenin. The seeds contain 0.1% alkaloids comprising the cardiac glycosides euonoside, euobioside and euomonoside. Euonymin has a cardiotonic action similar to the foxglove glycosides, but less intense, and it irritates the digestive tract.
**Symptoms** Ingestion of spindle-tree berries causes vomiting and diarrhoea, with colic and crampy pains. There is stimulation of the heart and mental confusion, with loss of consciousness. Onset of symptoms may be delayed for twelve to twenty-four hours after ingesting the fruit. In the experience of the NPIS, human symptoms of poisoning are usually mild, with no undue sequelae and no deaths reported. Cases of poisoning are frequent when the attractive fruits are ripe, and most often involve children. Animals will eat the leaves, which are said to be less toxic after the berries are formed, and poisoning is uncommon. Sheep and goats have been poisoned by eating leaves and twigs. Horses eating the leaves have suffered violent purgation. In one case, poisoning of two horses caused paralysis of the digestive tract and fatal impaction.
**Treatment** The patient should be made to vomit and then treated by gastric lavage. Demulcents and activated charcoal should be given orally, with fluid replacement and supportive care where necessary. Blood pressure and pulse rate should be checked, and heart function monitored by ECG.

# · BUCKTHORN ·

***Rhamnus catharticus* L.**

POISONOUS

*Fruit*
The fruit of buckthorn is a drupe in the shape of a flattened ovoid, about 8 mm broad and 6 mm high, with a small depressed central scar at the apex. When first formed, the fruits are green, turning brownish red and then purplish black and shiny. They are borne in short-stalked clusters on the old wood, and are readily detached when ripe in September and October. The skin is tough and leathery, and the fruit contains a greenish, juicy pulp, very slimy and sticky, with a pungent, not unagreeable odour and a bitter taste. Each fruit contains two to four relatively large pyrenes 4.5 × 2.5 mm, smooth, pale brown and sharply angled.

*Leaf and Flower*
Buckthorn is a tall deciduous shrub 3-6 m high, with wide-spreading branches and broad oval, finely toothed, opposite leaves which turn yellow and brown in autumn. The terminal leaf shoot is usually armed with a sharp thorn. Flowers are produced in May and June on ridged spurs on the old wood, which resemble roebuck's antlers – hence the name 'buck'shorn tree'. The flowers are usually unisexual, and in most – but not all – cases male and female flowers are borne on different plants. The flowers are tiny, 3-4 mm across, with four green petals, which are shorter and narrower in the female flowers. The flowers are strongly honey-scented, and clearly insect-pollinated, especially by honey bees. The process is efficient, and most bushes fruit heavily; one bush 2.1 m high was recorded as bearing 1,455 berries.

*Ecology and Distribution*
Buckthorn is widespread and frequent in thickets and hedges, especially on chalk and limestone in the south and east of England. In fenland it grows

with alder buckthorn (*Frangula alnus*), these two being the most important shrubs of alder carr. It grows on the limestones of Derbyshire and the Mendips, in sheltered places out of the wind. Elsewhere in the north it is uncommon on calcareous soils. It is probably not native in Scotland, and is absent from much of Wales and the south west of England. In Ireland it is present in a wide belt across the centre of the country including the Burren, but is absent from the south. In Europe it grows on calcareous soils in most countries north of 61°45′ in Sweden, and it is absent from the extreme south. It acts as the overwintering host of oat crown rust (*Puccinia coronata*), which develops on the young leaves in early spring, forming golden circles on the dark green leaves in May.

*History and Uses*

For centuries buckthorn berries have been used for their purgative properties, and syrup of buckthorn was prepared from them. An interesting light is thrown on mediaeval habits by the finding of buckthorn seeds in the drain of the reredorter of St Alban's Abbey. They were taken as a purgative, and must have been extensively used by the monks for so many to have remained for so long. The action of the berries is so violent that Dodoens in his Herbal commented 'They be not meat to be administered but to the young and lusty people of the country, which do set more store of their money than their lives'. Deakin wrote: 'The berries are powerfully cathartic, and were formerly used in medicine, but their action is very violent, and attended with so great pain, that now they are rarely used.' The ripe fruits are eaten by thrushes, blackbirds, waxwings and great tits, but only in severe winters.

*Toxicology*

The berries, both unripe and ripe, and the fresh bark are toxic, containing the anthracene glycosides rhamnosterin, rhamnicoside and rhamnicogenol, which on hydrolysis yield the purgatives anthrone and anthranol. Anthraquinone derivatives rhamnin

**Buckthorn**

and rhamnocathartin are present; also the flavone glycosides rhamnofluorine and frangula-emodin. These compounds cause vastly increased peristaltic movements, especially in the large intestine, and gastric irritation.

**Symptoms** Poisoning causes violent purgation, with abdominal pains and cramps, nausea and vomiting. There may be intestinal haemorrhaging, with the passage of watery bloody stools, and of alkaline urine, coloured red from the anthraquinone derivatives. Similar symptoms are produced in animals, with violent catharsis, prostration and coma. Overdosage of a dog with a laxative prepared from buckthorn has been known to cause death from severe gastrointestinal haemorrhages. NPIS have recorded no serious cases of poisoning in man, and it appears to be rare in all other species.

**Treatment** Vomiting should be induced if it has not already occurred, and gastric lavage using a demulcent and 5-10g activated charcoal. Dehydration and electrolyte imbalance should be corrected with fluid therapy, and antispasmodics used to control purgation and pain.

# · ALDER BUCKTHORN ·

***Frangula alnus* Miller**

POISONOUS

*Fruit*

The fruit of alder buckthorn is a drupe containing two to three pyrenes. The fruits are more rounded than those of buckthorn, 6-10mm in diameter, and carried in smaller clusters on longer pedicels from the leaf axils, so that the ripe fruits are pendulous. When first formed they are green, turning scarlet, and finally shiny purple-black when ripe from September to November. A few may persist on the bushes until December. The calyx remnants persist at the base of the fruit, contracted into a shield-like plate. The skin of the fruit is tough, and the sparse flesh is purple-staining. Each of the pyrenes is obovoid, 5 × 2mm, with a deep furrow at the base.

*Leaf and Flower*

Alder buckthorn is a deciduous shrub 3-5m tall, slenderer than buckthorn, with erect, thornless branches. The alternate, untoothed leaves, broadest above the middle, have a glossy upper surface and turn a brilliant yellow or red in the autumn. The flowers are small and green, with five petals 3mm across carried solitary or in small clusters in the leaf axils. They are hermaphrodite, with five stamens and a single nearly sessile stigma. They open in May and June, attracting numerous wasps, bees, flies and beetles, and flowering may continue into September as the fruits ripen.

*Ecology and Distribution*

Alder buckthorn is locally common in damp woods and thickets, especially on peaty acid soils, and with buckthorn it constitutes the major shrub element of alder carr in fen country. It requires moist, uncultivated soils, and is absent from much of central England, and from Pembrokeshire, Radnorshire, Montgomeryshire and Anglesey in Wales. It is a feature of some raised bogs, such as the mosses of Lancashire and Cheshire, the Somerset 'heaths', and on moist heaths, marginal scrub valley bogs and limestone scrub in the Lake District; less commonly in the west of Ireland. It is rare and possibly introduced in Scotland at Auldearn, Langholm, Dumfries and Newton Stewart, but was found in 1967 in two sites in Flanders Moss by the Glasgow to Aberfoyle road. Both these sites were very acid, with a pH of 3.5. It always avoids dry districts. In Europe it is widely distributed, except in the extreme north, and in much of the Mediterranean.

*History and Uses*

The bark of alder buckthorn was used in the past as a purgative, a dangerous practice since the bark must be stored for at least one year, otherwise it causes violent vomiting and colic. In 1871 Deakin wrote: 'The whole plant, but especially the bark, contains a bitter astringent principle, and has been used for the cure of ague and dropsy, and as a gargle in relaxed throats. But the principle use to which it is applied is for giving, by dyers, a yellow colour to woollen goods, and by the addition of some preparations of iron, a black colour is obtained.' The bark, leaves and unripe fruits yield a dye called sap-green.

Charcoal made from alder buckthorn was valued in making the best quality gunpowder, and the wood was known by the powder makers as dogwood. Before World War II most of the supplies used in Britain came from France and Czechoslovakia, and the Royal Ordnance Depot investigated the possibility of using Wicken Fen in Cambridgeshire as a source of supply. The fruits are eaten voraciously by fieldfares, and at Wicken Fen their roosts in the reed beds are liberally spattered with purple splodges. Field mice collect and store the seeds. A study showed that, of 1,804 fruits present on a bush on 27 September 1,268 had fallen by 24 December; of those 500 had disappeared, presumably eaten by birds.

*Toxicology*

The berries, leaves and inner bark contain the anthraquinone glycosides frangulin, glucofrangulin, frangula-emodin and frangularoside, also chrysophanic acid, an iso-emodin, tannic acid and bitter principles. The dried bark contains more than 6% hydroxyanthracene derivatives, with a higher proportion of frangulin than a fresh extract; consequently this has a more potent emetic action. The berries do not seem to be quite so potent. The action of the principles of alder buckthorn is very similar to those of buckthorn, causing gastric irritation and violent catharsis. They also stimulate bile secretion.

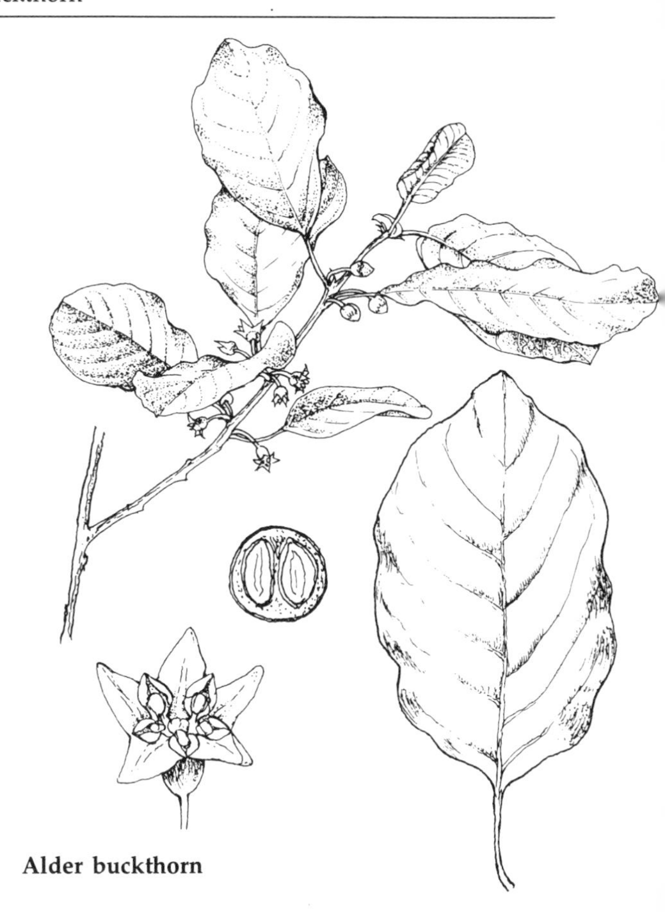

**Alder buckthorn**

**Symptoms** Poisoning causes nausea, vomiting and violent purgation, with colic and painful cramps, and the passage of watery bloody stools. NPIS report three human cases, none of them fatal. Poisoning in animals is rare, but in Sweden, in 1942 Södermark reported the poisoning of a cow which had eaten the leaves and fruits.

**Treatment** Vomiting should be induced, followed by gastric lavage and by the administration of demulcents. Fluid replacement therapy may be necessary, with the injection of antispasmodics such as dicyclomine or atropine. The patient should be kept warm and under observation for at least thirty-six hours.

# · MEZEREON ·

**_Daphne mezereum_ L.**

POISONOUS

*Fruit*
The fruit of mezereon is a drupe, borne in dense clusters close set to the stem on the one year-old wood. The fruits are ripe in early July, irregularly oval, 8-12mm long, broader near the top, with a shallow depressed scar at the apex. They are bright red and very glossy, with a thin skin and translucent orange flesh, which is very juicy with a slightly acrid smell and non-staining. The single seed, 6mm long and 5mm in diameter, is hard and globular with a pointed apex. When ripe it has a leathery, greenish black skin, brittle in texture. Plants with yellow fruits and white flowers (var.*alba* Aiton) are cultivated and occasionally naturalized.

*Leaf and Flower*
Mezereon is a small deciduous shrub up to 2m high, with thin, pale green, lanceolate leaves, mainly borne in whorls near the top of the stem on the new wood. When first opened they closely resemble the leaves of young privet (*Ligustrum vulgare*). The flowers are produced early, from late February to early April, opening with the leaf buds, unlike the garden form where they appear a week or two earlier and open before the leaf buds. The sessile flowers, borne in clusters of two to four are pinkish purple and sweetly scented. There are no petals, and the four perianth lobes form a distinct tube with four spreading points. There are eight stamens inserted at the top of the perianth tube.

*Ecology and Distribution*
Mezereon is a very rare and local plant of woodlands on limestone and chalk. It still exists in a few sites on the southern chalk from Wiltshire to East Sussex, in the Chilterns, in west Suffolk, and in the north on the limestones of West Yorkshire, south Lancashire, Staffordshire, Derbyshire and Cumbria. In 1985 it was discovered in the Ddol Uchaf nature reserve in Flintshire. In the south it has decreased as a result of collecting, and also because the woodlands where it used to flourish have

become overgrown, as less manpower is available to keep woods and rides clear. In Europe it is mainly a calcicole, growing widely except in the extreme west, south and north.

*History and Uses*
Despite their highly poisonous nature, in folk medicine the berries have been used against cancer, and as a purgative. Grieve noted that a dose of thirty berries is used for this purpose by Russian peasants, although French writers regard fifteen as a fatal dose! Birds have been recorded eating the berries: blackbirds have been seen taking them with avidity. They are often devoured by snails even before they are fully ripe.

*Toxicology*
All parts of mezereon are poisonous, especially the bark, root bark and berries. The fruits contain 0.04% of a vesicant resin, mezerine, and the bark contains a coumarine-like glycoside, daphnin, with coccognin and the aglycone dihydroxycoumarin. The berries are extremely poisonous and severely irritant to mucous membranes, causing gastroenteritis and renal damage.
**Symptoms** Human poisoning has resulted when the berries have been mistaken for red currants. One or two may be lethal for a child and twelve for an adult. Consumption of the berries produces violent diarrhoea with haemorrhage, and renal damage with haematuria. There is muscle twitching and a fall in blood pressure leading to collapse and death. They cause a burning sensation in the mouth and stomach, with swelling of the lips and tongue and profuse salivation. The irritation caused to mucous surfaces is intense, leading to blanching, sloughing and the production of a raw ulcer. Intact skin may be blistered by contact with juice from the berries. The literature cites many cases of human poisoning with severe symptoms, but the NPIS reports five cases where there was a severe burning sensation in the mouth and throat, but no eventual serious problems. Poisoning in animals is rare, as they are deterred by the irritant acrid juice, but there is a record of three berries causing death in a pig.
**Treatment** Medical attention must be sought immediately and the patient kept warm. **DO NOT INDUCE VOMITING WITH SALT WATER**. Give oral demulcents such as a mixture of eggs, milk and sugar, after which vomiting and gastric lavage may be induced, taking care not to cause further damage to the mucosal surfaces. Anaesthetizing salves may be needed for the skin lesions, with cortisone creams.

**Mezereon**

# · SPURGE-LAUREL ·

**_Daphne laureola_ L.**

POISONOUS

*Fruit*

The fruit of spurge-laurel is a drupe like that of mezereon, borne in similar fashion in tight clusters on the old wood just below the dense whorls of new foliage at the stem tip. The berries are juicy, purplish black, about 10mm long, and 6mm in diameter, pointed oval with a small depression at the apex. They contain a single very hard seed, 7mm long and 3.5mm wide, with a dark green, shiny and refractile surface. The basal end is blunt and the free end sharply pointed. The fruits ripen in late June and July, but seldom remain on the plant for long.

*Leaf and Flower*

Spurge-laurel is a small evergreen shrub 0.5-1m high. The leaves are pointed lanceolate, narrow based and leathery, with a dark green shiny surface. They grow in whorls on the new wood at the stem tips, leaving the pale brown stems below almost denuded of foliage. The flowers, which appear early in the year from late January to April, are small and yellow-green, in dense axillary clusters of five to ten on the upper part of the stem below the new leaves, which almost conceal them. Each flower has a spoon-shaped bract one third of the length of the tubular flower, which is formed of four fused perianth segments with spreading, pointed lobes. There are four stamens with orange anthers visible at the throat of the tube. The flowers are faintly scented.

*Ecology and Distribution*

Spurge-laurel is fairly common in dry woods and hedgerows where it is easily overlooked. It occurs on basic soils in south England, the Channel Is-

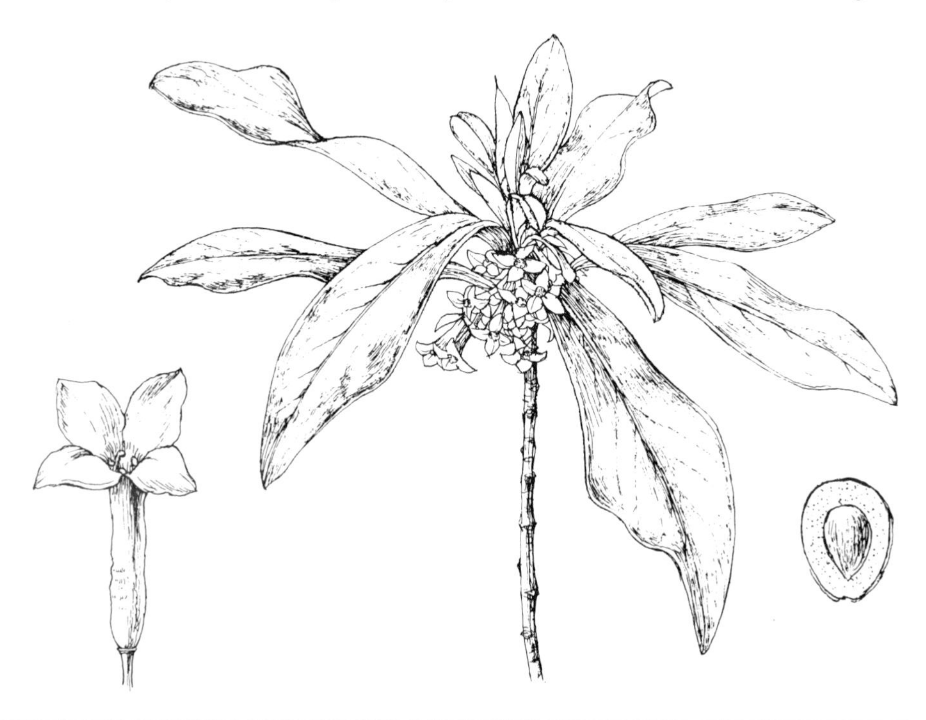

**Spurge-laurel**

lands and Wales. In south and south east England it is a characteristic plant of beechwoods on chalk, flourishing in bare areas of dense shade, since it flowers early in the year when light values beneath the leafless beech trees are adequate. It grows as far north as Cumbria and there is a site on limestone near Appin in Argyllshire. It has also been recorded recently from Wigtownshire and Roxburghshire. Otherwise it is not apparently native in Scotland and it is not native in Ireland. In Europe it grows in the south, south central and western regions, north to Hungary.

The ripe fruits rarely remain on the bushes for long. G.H. Knight found that plentiful berries were set by May, but by mid-June all had been removed, even in the unripe state. He suspected bank voles and woodmice, for he found 'the berries neatly removed, skins peeled off and left lying, the stones opened and seeds eaten'.

*History and Uses*

Spurge-laurel has scarcely been used medicinally on account of its toxicity, but Deakin noted that 'The bark has a hot, pungent taste, and is a powerful stimulant, and often useful when made into a gargle for relaxed throats'.

*Toxicology*

All parts of the plant are poisonous, especially the bark and berries. They contain the coumarin glycoside daphnin and the vesicant resin mezerine which cause severe poisoning, with a burning sensation in the mouth and stomach and swelling of the tongue and lips, followed by gastro-enteritis. **Symptoms** Consumption of the berries causes blistering of the mouth and tongue followed by diarrhoea with haemorrhage, delirium and collapse. The berries are attractive to children, but as they are far more acrid than the fruits of *D.mezereum*, livestock rarely touch them. Horses have been poisoned with them, showing symptoms of intense colic and constipation, followed by superpurgation with blood and shreds of mucous membrane in the faeces. Recently some calves were poisoned near Leeds. One died and two showed severe colic.

**Treatment DO NOT INDUCE VOMITING WITH SALT WATER**. Give oral demulcents such as a mixture of eggs, milk and sugar, or vegetable oil, and seek medical attention immediately.

# · SEA-BUCKTHORN ·

***Hippophaë rhamnoides* L.**

*Fruit*

The fruit of sea-buckthorn is a 'berry' with a very curious structure. It is an achene within a membranous ovary wall, the whole enclosed in a fleshy receptacle. The succulent orange berries, 8-13mm long, are borne on very short stalks on the female plants, the fruits densely clustered as many as one hundred on a 10cm length of stem. They start to colour up in early August and may remain on the bushes until the following February or March. The skin of the berry is smooth, liberally covered in tiny brown scab-like scales which are concentrated mainly at the base and apex of the berry. The flesh is thick and juicy, with a pleasant fruity smell and pungent taste. Each fruit contains a single shiny brown, ovoid seed, 4 × 2.5mm, with a pointed apex and a deep sulcus running up each side of the seed.

*Leaf and Flower*

Sea-buckthorn is a deciduous, thorny shrub, 1-9m tall, with an extensive horizontal root system which leads to the formation of dense thickets. The thorns form on both terminal and lateral shoots, and are long, straight and formidably sharp. The leaves are narrow lanceolate, alternate, almost sessile and covered on both surfaces with silvery scales. Some plants may have leaves which are glabrous above or brown beneath. Sea-buckthorn is dioecious: the tiny green male flowers, 4mm across, form on the year-old wood in small spikes of four to eight flowers in the leaf axils, appearing from mid-March to late April, before the leaves open. There are two perianth lobes, no petals, and four stamens. The female flowers, 1.5mm across, are solitary or in short racemes, opening with the first leaves. The perianth is tubular, with two tiny lobes and a large protruding style. Pollen is wind-carried – no problem in the exposed coastal sites where it grows.

*Ecology and Distribution*

Sea-buckthorn occurs widely as a native plant on

the sea coast of eastern Britain, from Berwick in the north to Camber on the Sussex coast in the south. It is particularly common in areas of sand hills, more rarely on sea cliffs, and has been introduced in many sites around the coasts of south and south west England, Wales, Scotland and also Ireland, mainly in the north east and south east. It has been recorded from a few inland sites in southern England and in Scotland from Dumfries and Roxburghshire. On the Norfolk coast at Hemsby and Wolferton it grows in profusion. In the South Lancashire dunes between Southport and Formby it was introduced around 1883, and now forms dense and extensive thickets. In Carmarthen it now grows on both sides of the Tywy estuary at Pendine and Tywyn Burrows – all evidence of its ability to colonize successfully any suitable habitat. Well established bushes can grow as much as 700mm in one year, and the rhizome-like roots establish new plants quickly, growing horizontally at 5m yearly in open sand. In exposed coastal areas the bushes provide useful cover and feed for migrating passerine birds, but they can easily get out of hand and cause conservation problems. Since the bushes are dioecious the process of rapid vegetative reproduction can lead to the production of unisexual colonies. Sea-buckthorn is widespread in Europe on stable dunes, sea cliffs and on river gravels in mountains from 68°N in Norway and in Finland, south to central Italy, north west France and northern Spain.

Two subspecies occur in Asia. Ssp.*thibetana* (Schlect)Servattaz is a dwarf plant, sprouting annually from basal shoots. It occurs in Tibet at 2300-5000m. Ssp.*salicifolia* (Don)Servattaz grows in the more temperate Himalaya from 1600-3500m. It is a tall shrub with longer, narrower leaves and less spiny branches. It has been seen at 3500m growing in profusion at Panikhar in the Suru Valley in Ladakh, where the branches are used to render fences stock-proof. Our own plant can be assigned to ssp.*maritima* Van Soest, having many thorns, short stiff branches, and a compact inflorescence.

*Uses*

The fruits contain large quantities of vitamin C, carotenes, fruit acids and a fatty oil. The fresh juice can be preserved with honey and used as a tonic. In countries bordering the Gulf of Bothnia the berries are used to make a jelly, which is eaten with fish.

**Sea-buckthorn**

The consumption of the berries in large quantities is not to be recommended, as they tend to be purgative, and have been used in the past as a vermifuge. Birds will eat them when other food is in short supply; in south Lancashire predation has been recorded by fieldfares, blackbirds, redwings, mistle thrushes, ring ouzels, bullfinches, waxwings and even an overwintering blackcap. When the bushes are stripped the birds still use them as a roost, although they depart daily to feed inland. Redwings and fieldfares have been observed coughing up and ejecting the less digestible skin and seed. Hooded crows have been seen eating the berries on the island of Mull.

# · TUTSAN ·

**Hypericum androsaemum L.**

? POISONOUS

*Fruit*

Tutsan is unique among our native Hypericums in having a berry as a fruit. The fruits are borne in flat topped sprays of two to eight at the top of the stems above a pair of pointed leaves, which are almost conjoined at the base. One plant may carry as many as 120 fruits. Each fruit, 9mm long and 7mm in diameter, is pointed oval with two to three stylar remnants at the tip, and usually five broad reflexed green sepals, flecked with red, like a jester's collar at the base. The berries ripen from green to bright red, and finally purplish-black in late August and September. The hard, thick-walled berry is divided by three shallow vertical grooves into three loculi. When ripe the skin is brittle and the berry splits easily into three parts, each containing numerous dark brown oval seeds about 1mm long. It is estimated that each fruit contains eight to nine hundred seeds. The ripe fruit is slightly aromatic.

*Leaf and Flower*

Tutsan is a hairless, half-evergreen, low shrub 30cm-1m high. The reddish stem is winged and has large, sessile, half-evergreen, pointed oval leaves, 50-100mm long, in opposite pairs; they are aromatic when crushed. The yellow flowers 12-20mm across, open from June to August in flat-topped sprays of two to eight. The five petals just exceed the blunt oval, unequal sepals, which reflex and flush with red as the fruit ripens.

**Tutsan**

*Ecology and Distribution*
Tutsan has a slightly westerly distribution, being widespread but local in south west England, Wales, north west Scotland and throughout Ireland. It is rare or absent from many of the eastern counties of England and Scotland. It is a plant of wood margins, hedges and cliffs on base rich soils, often growing near streams, and tolerating quite heavy shade. In north England in some areas of limestone pavement it grows deep in the sheltered grikes between the limestone blocks. In Europe it grows in damp, shady places mainly in the west, locally in southern Europe, and east as far as Turkey.

*History and Uses*
Hypericum was associated with St John the Baptist as early as the 6th Century, and may have been used in the Midsummer festival in pre-Christian times. Its use to repel evil spirits and witches was widespread in the British Isles and Europe. The specific name *androsaemum* derives from the Greek *andros* (man) and *aima* (blood), and refers to the red stain which develops on the leaves and sepals and which is supposed to be symbolic of the blood of St John the Baptist. Hampshire tradition holds that the berries are stained with the blood of the Danes.

Deakin wrote: 'The pulpy fruit was formerly used, when bruised into the form of a poultice, to dress recent wounds, and from its healing properties it obtained the French name of Toute-saine, and the English name Tutsan is probably derived from that'. Culpeper used the leaves: 'It purges choleric humours, helps the sciatica and gout, and heals burns.' In Dorset it was called Book-leaf or Bible-leaf, the aromatic leaves being placed between the pages of the family Bible for their pleasant perfume.

In man, St John's-wort taken internally is used as a spasmolytic, is mildly diuretic, and stimulates gastrointestinal secretions especially bile. It has a proven action as a cicatrizing agent on wounds, cuts and bruises, and helps reduce the size of haematomas. It is used in homeopathic and naturopathic medications.

*Toxicology*
The toxicology of tutsan is not well documented, most of the work having been done on *H.perforatum*. This contains a glycoside hypericin, which is a red pigment, a volatile red oil, a polyphenolic flavonoid hyperoside, tannin and carotene. Hypericin is not destroyed by drying in hay or by storage. When eaten by animals it leads to photosensitization in areas of skin unprotected by the pigment melanin; the ultraviolet component of sunlight then causes severe burning of the stratum germinativum of the skin, and white-coloured areas necrose and dry.

**Symptoms** NPIS record two human cases of poisoning, which were asymptomatic. Animal intoxication is well documented and frequent, most cases referring to *H.perforatum* and *H.elodes*. There is no record of poisoning by tutsan. Photosensitization lesions in animals can be very severe; muzzles, eyelids and teats become as hard as wood, and whole areas of white-coloured skin peel off like wallpaper. Affected animals may well have to be slaughtered on humane grounds, since affected lactating animals cannot be milked.

**Treatment** In the early stages, injections of antihistamines and steroids can be of help, but established lesions take months to heal, during which time they must be dressed with an emollient, fly-repellant cream.

# · *WHITE BRYONY* ·

***Bryonia cretica* L.subsp. *dioica* (Jacquin) Tutin**

POISONOUS

*Fruit*

The berries of white bryony are borne in drooping clusters on the shrivelled remains of the climbing stems, for by the time the fruits ripen, from July to October, the leaves, stems and coiled spring-like tendrils have usually withered. The berries are spherical, 5-8mm in diameter, first green, then yellow, ripening to pinkish-red with a dull skin, and usually carried in groups of two or three in the leaf axils, on very short, fine pedicels. The berries are soft and squashy, with orange-brown juice which has an acrid and very clinging smell when handled. Each fruit is divided into three loculi, each containing two ovules, but usually only three to four seeds mature. The seeds, 4.5 ×3.5mm, are smooth and fat, pointed oval and marked with black on a pale brown base, like a lapwing's egg.

*Leaf and Flower*

White bryony is the only native British representative of the marrow family, *Cucurbitaceae*, and is a robust, climbing perennial, growing annually to 4m from a massive, turnip-shaped tuber. The stem is rough, tending to climb anti-clockwise, aided by unbranched axillary tendrils coiled like springs. The alternate leaves are palmate, with five broad lobes and a rough surface like sandpaper. White bryony flowers from late May to early September, the flowers usually unisexual and carried on different plants. The male flowers, 20mm across, have five sepals and five greenish-yellow petals with dark veins, and are carried several together in a stalked raceme from the axilla of the leaf. There are five stamens, two pairs united by their filaments, and one free. The female flowers, 10mm across, are borne in sessile pairs, and have three prominent bifid stigmas.

**White bryony**

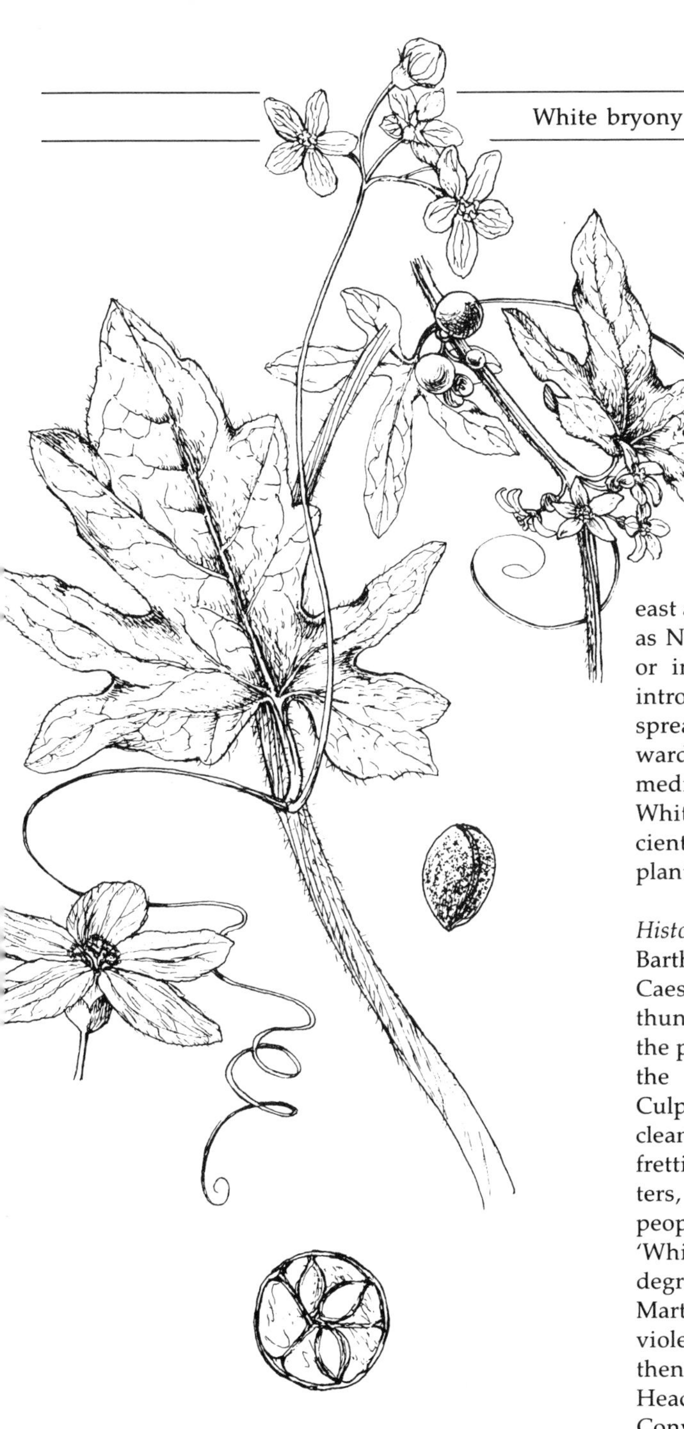

## *Ecology and Distribution*

White bryony is common in hedges and scrub in the South of England and the Midlands, especially on the chalk and greensand. In south west England it does not grow as a native further west than Somerset, and in Wales grows mainly in the south east and north east. It grows in England as far north as Northumberland, but it is not native in Ireland or in Scotland, where there are two records of introduced plants in Angus. In Europe it is widespread in the south, south centre and west, northwards to Britain. Formerly it was cultivated as a medicinal plant from which it is often naturalized. White bryony can be pollinated by insects of sufficient size, such as Apids, and bryony is the food plant preferred by the oligophagic *Andrena florea*.

## *History and Uses*

Bartholomew's *Anglicus* records that Augustus Caesar used to wear a wreath of bryony during thunderstorms to protect himself from lightning. In the past the violently purgative root was the part of the plant most commonly used in medicine. Culpeper wrote: 'The leaves, fruit and root do cleanse old and filthy sores, and are good against all fretting and running cankers, gangrenes and tetters, and therefore the berries are by some country-people called tetter berries'. Gent in 1681 wrote: 'White Briony is hot and dry in all parts in the third degree, both the white and the black are furious Martial Plants, and purge the Belly with great violence, and therefore you are to correct it, and then it is very good for all manner of Griefs in the Head, as also the Joynts and Sinews, Cramps and Convulsions, Dropsey, provoketh Urine, and is good for the Stone.' Tincture Bryonia was a popular remedy for diseases of the liver and spleen. In France white bryony was called Navet du Diable (Devil's Turnip), and has caused poisoning in women who have eaten it in the mistaken belief that it will reduce milk secretion at weaning time.

*Toxicology*
The whole plant contains an acrid, milky juice, nauseous when dry and irritant to the skin. The root and berries are the most poisonous part and contain the glycosides bryonin and bryonidin which are drastically purgative. In addition the seeds contain a saponin and the fruit contains the dye lycopine. The berries are a gastric and mucosal irritant, poisoning producing symptoms of acute gastro-enteritis, with vomiting, colic and purgation. Children have been known to eat the berries, fifteen proving a fatal dose, while more than thirty may prove fatal to an adult. The roots have on occasion been mistaken for turnips. Poultry will eat the berries, and pigs and cattle have been known to eat roots exposed by digging and drainage work. Animals usually leave it alone, but cows can become addicted to it, and will break through fences to get back to a supply. A case of poisoning in a border collie, which ate the berries with fatal consequences, was reported in 1986.

**Symptoms** The berries are irritant to the lips, tongue and mucous membranes, causing oedema of the oesophagus. There is nausea, vomiting and protracted diarrhoea, with abdominal cramps and the passage of watery, blood-stained stools. Fluid loss leads to dehydration. NPIS have recorded seven cases, all of which exhibited mild symptoms, with no sequelae and no deaths. Poisoning in cattle produces symptoms of collapse, the animal being cold to the touch with deeply sunken eyes and a profuse watery scour. The poisoned collie dog showed initial symptoms of elevated temperature, and increased pulse and respiratory rate. This was followed by sudden collapse and death within 24 hours. A post-mortem examination revealed extensive internal haemorrhages, and seeds identical to those within uneaten berries were recovered from the intestinal contents.

**Treatment** Vomiting should be induced if it has not already occurred, followed by gastric lavage using demulcents and the administration of activated charcoal by mouth. **TANNIC ACID SHOULD NOT BE USED AS A LAVAGE**. Fluid replacement therapy may be needed to correct electrolyte imbalance, and supportive care continued over twenty-four hours. The airway should be checked repeatedly, since swelling of the tongue and oesophagus can occlude it.

Intravenous fluids and oral demulcents must also be used in cases of poisoning in animals, where dehydration can be a problem.

# · FUCHSIA ·

**_Fuchsia magellanica_ Lam.**

Fuchsia exists in a vast number of cultivars, many of which are not hardy, but some have established themselves widely and have been used for hedging, especially in the more oceanic climate of the west coasts of the British Isles. Two types of fuchsia are widespread in Ireland and elsewhere, and the following descriptions are based on a paper by E.C. Nelson and personal observation.

## *Fuchsia magellanica* Lam.sensu stricto

*Fruit*
The fruits of fuchsia are four-celled, loculicidal capsules, and are ripe from August until the first frosts. The pedicel is long and held parallel to the main stem, bending at the tip to the single pendulous fruit.

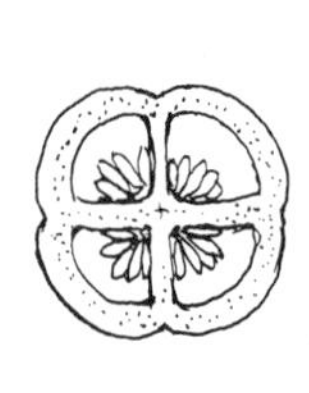

***Fuchsia magellanica***

**Fuchsia**

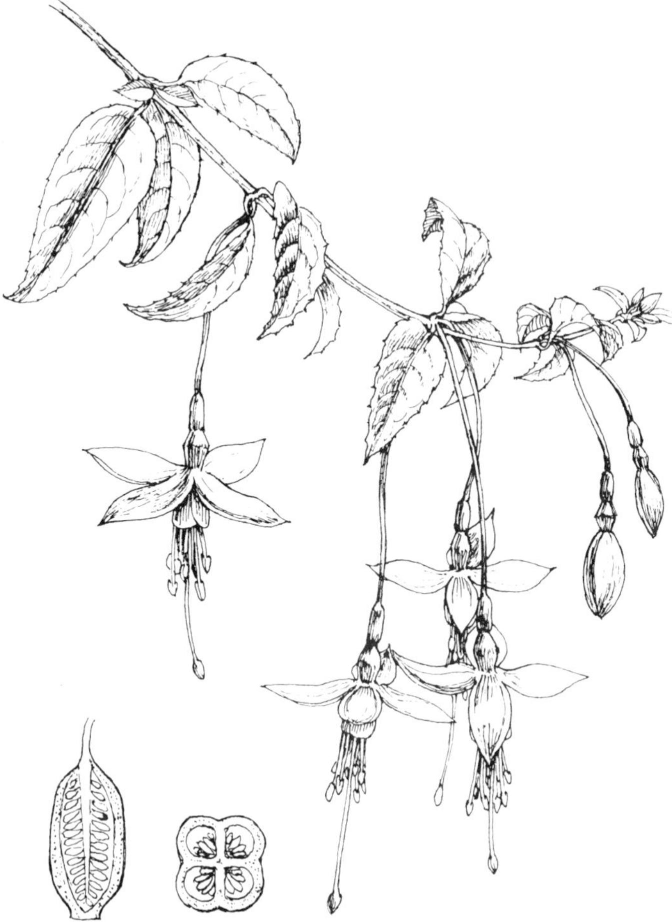

***Var. Riccartoni***

The fruit of this type is 12mm long, 5mm wide, plump, burgundy-coloured and shaped rather like a banana. The apex is pinched into a small flat scar. The fruit, which is produced sparsely, is juicy, the juice being purple-staining, and there are numerous tiny yellow seeds about 1mm long, flat and triquetrous. The seeds are usually viable.

*Leaf and Flower*
Fuchsia is a deciduous shrub 1-3m high with opposite, toothed, pointed oval leaves which are often tinged red when young. The distinctive, solitary, pendulous flowers are produced from June to November. In this type the four long pointed, turkey-red sepals are 4mm broad and parallel-sided, enclosing four plum-coloured, broad oval petals which overlap in the shape of a cup. There is a prominent stigma, slightly curved and club-ended, and eight red stamens, four long and four short. The shape of the mature flower-buds is important for identification; they are long and thin, 17mm from the attachment of the pedicel to the tip, and 6mm wide at the broadest point. This type is rare in Ireland.

## *Fuchsia magellanica* Var.Riccartoni

*Fruit*
The fruits of this type tend to drop before maturity, but when formed they are oblong, about 9 × 4mm,

with a broad flat apex, mauve-purple marked with numerous small rough green spots. They contain numerous small seeds, the viability of which is not known – var.Riccartoni may not be fully fertile.

*Leaf and Flower*
Growth is the same as that of *F.magellanica* sensu stricto, but this type is distinguished by its fat, squat flower buds, which when mature are 17mm long and 10mm broad at the widest place. The red sepals are broad lanceolate-ovate, 7-9mm in width. This form is more frequent in Ireland.

*Ecology and Distribution*
Fuchsia is usually found in hedges. In many western areas of Ireland and in Cornwall it was planted extensively as a hedge plant on the stone walls, looking most attractive when bordering a lane, with the blue sea beyond. In Ireland it is extensively naturalized from Co Clare round the south west coast to Waterford. It is widespread in south west England and has flourished in the warm, moist climate of the Isle of Man. It is an established alien in hedges in Wales – in Glamorganshire, Pembrokeshire, Cardiganshire, Merioneth, Caernarvonshire and Denbighshire. It also grows on the moors in Orkney and in many places in the south and west of Mull. In the Azores it was planted for hedges.

*History and Uses*
Fuchsia was named after the 16th Century German physician and herbalist Fuchs. It was introduced into Britain from the Falkland Islands in 1823 and welcomed as a highly decorative shrub, with some value as hedging material. There is no evidence that the fruits are poisonous.

# · *DOGWOOD* ·

**_Cornus sanguinea_ L.**

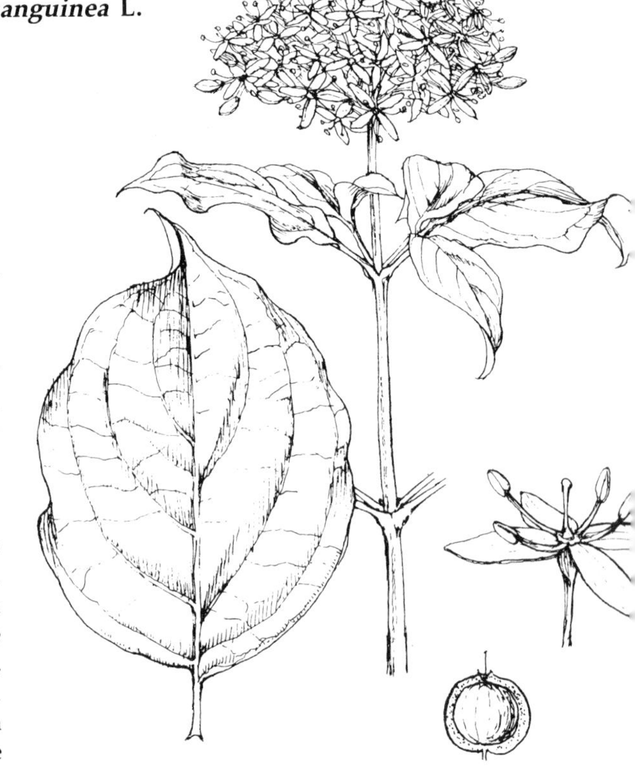

*Fruit*
The fruit of dogwood is a black, globular drupe 5-8mm in diameter, carried in flat-topped corymbs of ten to fifteen on bright red pedicels. The skin of the fruit is fairly shiny but dotted with minute rough cells. The apex of the fruit bears the scar of the calyx remnants. The fruits are juicy, with reddish-staining juice and a bitter astringent taste. They contain globular nutlets 3mm in diameter which have a smooth surface marked with a vertical furrow. The fruits are ripe in September and October.

*Leaf and Flower*
Dogwood is a small, spreading, deciduous shrub 1-4m high, with opposite, pointed oval leaves, slightly downy on both sides, and marked with three to five pairs of conspicuous curved veins. The leaves turn a most attractive pinkish red in autumn, and after they fall they reveal the coral red twigs. The flowers appear from late May until July, with dense, flat-topped erect corymbs of creamy white

**Dogwood**

flowers. There are four tiny sepals and the corolla is four lobed, with narrow, acute petals, four stamens alternating with the petals, and a club-shaped style. The flowers have a faintly foetid scent.

*Ecology and Distribution*
Dogwood is abundant in the southern half of England and east Wales, showing a decidedly south eastern pattern of distribution. It is one of the major constituents of scrub on chalk and limestone and is plentiful in hedges and open woodland on calcareous soils. It is uncommon in south-west England. In the north it grows as far as Morecambe Bay and the Tyne. In Scotland it only exists as an introduced species, and in Ireland it is a rare plant in the southern half of the country, with a new record for Waterford in 1984. It grows in most of Europe except the north east and the extreme north.

*History and Uses*
Dogwood was also known as Dogberry and Hound's Tree, from the reputed efficacy of an infusion of bark, leaves and fruit as a cure for mange in dogs. In mediaeval times the fruits were used as a mild laxative. Grieve wrongly attributed 'Gaitrys beryis' as mentioned in Chaucer's Nun's Priest's Tale to *Viburnum opulus*. Pertelote, the hen, recommends them to her husband Chauntecleer–

> A day or two ye shul have digestyves
> Of worms, er ye take youre laxatyves
> Of lawriol . . . . . . . . . .
> . . . . . . . . . or of gaitrys beryis
> . . . . . . . . . . . . . . . . . . .
> Pekke hem up right as the growe and ete hem yn.

*Toxicology*
Cases of dermatitis in man have been attributed to contact with dogwood leaves.

# · DWARF CORNEL ·

***Cornus suecica* L.**

*Fruit*

The fruit of dwarf cornel is a brilliant, shiny-red drupe, 4-10mm in diameter, slightly irregular in shape, with the black calyx remnants at the apex. It does not fruit very well in the British Isles, so that few plants produce more than five ripe berries, but they are relatively large and eye-catching. The fruits ripen in the last week of August and throughout September. They are carried on very short pedicels in clusters above the leaves, which by then have attained their pinkish-mauve autumn colours. Mixed with the ripe fruits are the black, spiky remnants of the flowers which have not formed berries. The berry has a thin red skin and is filled with white flesh of a meringue-like consistency and sweetish taste. At the base of the fruit there is a single ovoid stone 4 × 3mm, pale brown and faintly pitted.

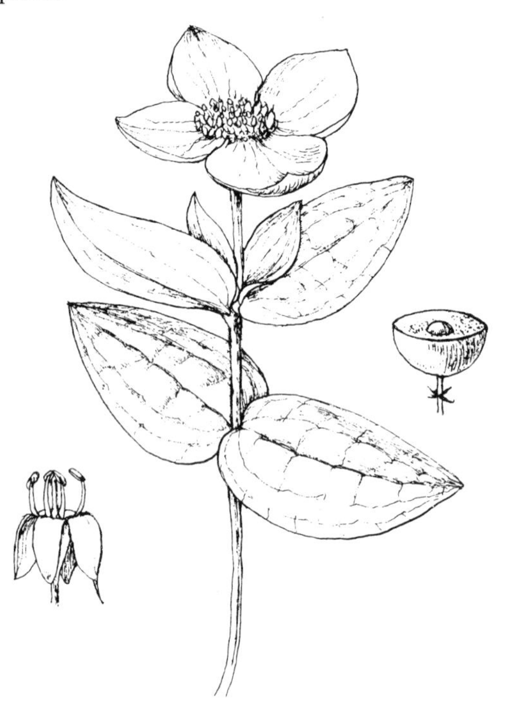

*Leaf and Flower*

Dwarf cornel is a small, creeping perennial with a simple herbaceous stem up to 25cm high. It is often concealed by the heather and bilberry with which it grows. The leaves are pale green, pointed oval, in unstalked opposite pairs, turning shades of red and mauve in autumn, which makes the presence of non-flowering plants much easier to detect. Flowers are produced from mid-June to late August in a small terminal umbel of eight to twenty-five. Each true flower in the umbel is very small, 2mm across and purplish black, set in groups above four large, white involucral bracts which resemble petals. In Norway and Sweden, plants with yellow-green flowers, var.*chlorantha,* occur rarely, but have berries of the normal red colour.

*Ecology and Distribution*

Dwarf cornel is widespread but overlooked on heathy moors and mountains in Scotland, mainly in the western Highlands, and in the north of England. It is usually to be found on acid rocks up to 900m, and is part of the arctic-alpine flora. It grows as far south as Turton Moor near Darwen in Lancashire, and Hole of Horcum near Saltersgate in north east Yorkshire. It is also found on the Cheviot and Rimside Moor in Northumberland. It does not grow in Ireland. It is a plant of the edge of the heather and bilberry zone on mountains, just above woodland margins, often in areas with scattered birches and moss covered boulders, growing with chickweed wintergreen, cloudberry and common cow-wheat, and it can easily be overlooked since it neither flowers nor fruits well in Britain. Elsewhere in Europe it is calcifuge, growing extensively in Norway and Sweden, south to the Netherlands. Its southernmost station is in pine and spruce plantations near Wilhelmshaven in north west Germany.

*History and Uses*
It was first mentioned in 1671 by John Ray, who found it in Westmorland, and then by Lightfoot who described it in 1772 growing near Little Loch Broom.

The berries are sweet but somewhat insipid, very pleasant when stewed with blueberries and other mountain berries. The Eskimos use them as a winter food. In Scotland they were known as lus-a-chraois – plant of gluttony – because of their tonic effect upon the appetite. Birds will eat them, and their Swedish name is 'Hönsbär' – chicken berry.

**Dwarf cornel**

# · *IVY* ·

***Hedera helix*** **L.** AND ***H.hibernica*** **(Kirch.) Bean**

POISONOUS

Recent work has shown that there are two ivy species native to the British Isles, which can be separated by careful examination of the shape of the scale hairs on the tips of the juvenile growing-shoots or very young leaves. It requires a high-powered hand lens or a microscope to see the scale hairs, which are star-shaped with rayed arms; they must be handled with care as they are very easily dislodged. A number of exotic ivies have established themselves locally, and the various species fall into two clearly defined groups:

1 Ivies from north and west of the Mediterranean and the Black Sea. The shoot-tips and young leaves bear scale hairs with long white rays. The species include *Hedera helix, H.hibernica* and *H.azorica* which is wild in the Azores.
2 Ivies from south and east of the Mediterranean and the Black Sea. The scale hairs of these species are red, and look much like red spider mites.

*Fruit*
The fruit of ivy is a five-celled drupe, although the full potential is rarely achieved, and most fruits contain fewer than five stones. The fruits are carried in a rounded umbel of ten to twenty on pedicels about 8mm long, which radiate in a sphere from the top of the fruiting stalk, with a few sometimes borne on the peduncle below. Each is 8-10mm in diameter, roughly spherical with a flat disc on top and covered with a dull, blackish-purple skin. The edge of the disc carries five scars, and the stylar remnant protrudes from the centre. The flesh of the fruit is dry-textured and greenish, with a bitter nauseous taste. Up to five stones may mature, 6.5 × 5mm, yellowish at first and wrinkled, resembling a small cowrie shell. The number of berries on each umbel varies greatly, and not all flowers mature to form fruits. The more exposed blooms appear to be more efficiently pollinated, and their fruits contain more seeds. The fruits start to ripen from November onwards and are usually at

**Ivy**

their best in December, remaining on the bushes until April if not eaten by birds. Plants with dark, glossy-green, ripe fruits occur, and a yellow-fruited ivy is known from Greece and Turkey and is sometimes cultivated in France, Italy and more rarely in Britain. This form is tetraploid, and thus is thought to be a low anthocyanin form of *H.hibernica*.

## *Hedera Helix*

### *Leaf and Flower*

Ivy is a woody, evergreen climber, growing on trees and cliffs to 30m with the aid of a multitude of adhesive roots or trailing extensively over buildings, rocks or woodland floors. The leaves of the non-flowering branches are five-lobed, leathery and dark green, while the leaves of the specially developed arboreal flowering branches are spear-shaped with wavy edges. The leaves are never large. The scale hairs are pale with six to ten rays, which radiate from the surface in all directions, like the spines of a hedgehog.

The flowers are yellowish-green, 6-10mm across, borne in rounded umbels of ten to twenty. They have five small tooth-like sepals, five recurved petals and five stamens with long, erect, yellow anthers. The flowers, which open from mid-September to mid-November, smell like elder flowers, and have a plentiful supply of nectar which attracts pollinating flies and wasps. *H.helix* is diploid – 2n = 48.

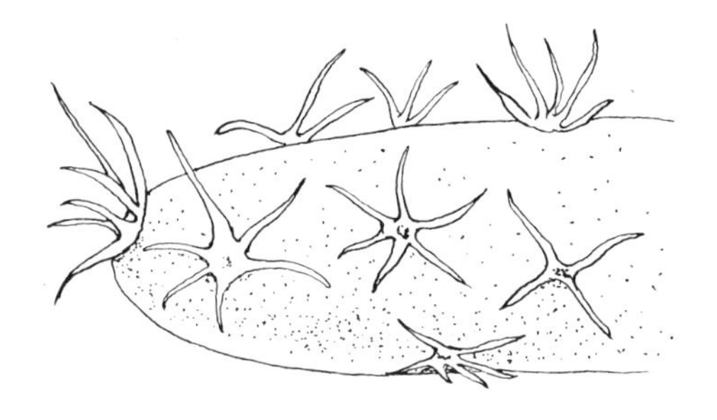

### *Ecology and Distribution*

*H.helix* is the most common wild ivy in Britain, growing in hedges, on trees, cliffs, old buildings and the floors of woodlands, from sea-level to 700m. It avoids very acid soils and extremes of wet or dryness. It is widely distributed throughout the eastern half of England and most of Scotland, but is less common in the West Country west of Hampshire, west Wales, the Isle of Man and Ireland. In October ivy flowers are a major source of nectar for wasps, flesh flies and green bottles (*Lucilia* sp.). Red Admirals, Painted Ladies, Small Tortoiseshells and honeybees are all frequent visitors to the nectar, which is fully exposed and abundant. From November to December there is a succession of flowers, which are important to the survival of queen wasps. The berries are particularly favoured by blackbirds, and are also eaten by song thrushes, mistle thrushes, redwings, wood pigeons and collared doves. Song thrushes have been observed avidly devouring ripe berries in April, when there was no shortage of alternative food. The seeds are distributed by birds, and they germinate in the following spring. First-year seedlings progress to two cotyledons, adding two foliage leaves in the second year, but they are tough and resistant to frost and drought. They grow best on slopes where leaf litter does not accumulate, and prevent the development of adventitious roots. Shoots can grow 60 cm in a year, but increases of 15 cm on the ground and 20 cm up trees are more usual. The development of flowers and fruit will depend upon the plant finding a suitable support, and the shoots can easily be torn off smooth-barked trees.

## *Hedera hibernicia*

*Leaf and Flower*
Atlantic ivy is very variable but is usually lighter olive-green in colour, glossier leaved and faster growing than *H.helix*. Fruits and flowers appear to be indistinguishable from those of *H.helix*, though fruiting is never as frequent or as heavy. The scale hairs are buff-coloured, with numerous rays spreading from a reddish-brown centre. The important distinguishing feature is that the rays lie parallel to the surface of the shoot or leaf, looking remarkably like a stranded sea-anemone. Atlantic ivy is tetraploid – 2n = 96.

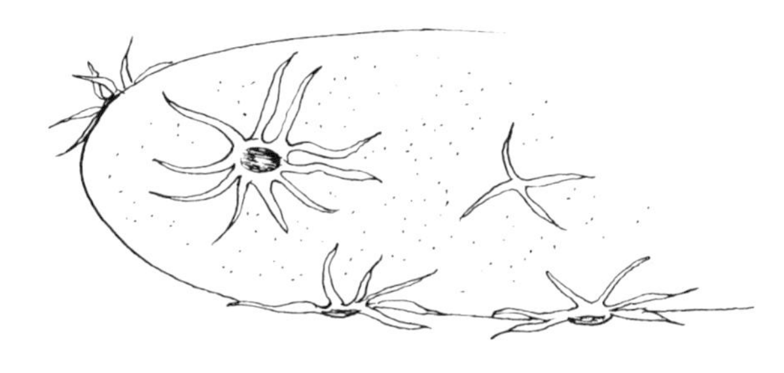

*Ecology and Distribution*
Atlantic ivy grows in similar situations to ivy. It is widespread in south west England and the Channel Islands and exists in mixed populations with ivy in Wales, Mull, south west Scotland and north Ireland.

In Eire, only *H.hibernica* will be found. This is also found on the Atlantic coast of Europe.

Garden 'Irish' ivy, *H.hibernica* c.v.*hibernica* is a sub-arboreal clone of Atlantic ivy, intermediate between the creeping and flowering forms, with its origin in the Isle of Wight and south west England. It has uniformly-sized, dark green glossy foliage which does not open flat, so that the sunlight tends to reflect from one half of the leaf. The leaves are broader than long, with wide short lobes and long, flesh-coloured petioles. It is a common ivy on garden boundary walls, in cemeteries and in woods – but *not* in Ireland! Much of it does not bear fruit.

## *Exotic Ivies*

*Persian ivy* (*H.colchis*) has deep, rich-green, glossy leaves with the surface raised between the main veins, giving a quilted effect. When warm it smells like celery. Some cultivars have grey and cream variegated leaves which are dull and leathery, with marginal prickles. The leaves tend to roll up in cold weather. The scale hairs are squat, with numerous orange-red rays.

*Algerian ivy* (*H.algeriensis*) The cultivar has large, shield-shaped leaves, variegated with cream splashes. The petioles and new bark are ruby red, and the cut stems smell like turpentine.

*History and Uses*
Ivy has been used medicinally for a long time, and yellow-fruited forms may have had ritual significance. Ivy was supposed to have magical properties to keep evil away from animals. Gerard wrote: 'The leaves laid in steepe in water for a day and a nights space, helpe sore and smarting waterish eies, if they be bathed and washed with the water wherein they have beene infused'. Culpeper noted: 'The yellow berries are good against the jaundice and a drunken surfeit . . . The berries prevent and heal the plague. The juice of the berries or leaves stuffed up the nose, purges the head and brain of thin rheum . . .' These sentiments were echoed by Gent in 1681.

An infusion of ivy can be used to remove shiny patches on clothes. The leaves of ivy are still used as a tonic when fed in small quantities to sheep and cattle. Honey made by bees from the nectar is pale amber coloured. No transfer of toxins to the honey is known.

*Toxicology*
The entire plant is poisonous to man, the toxic principles being concentrated in the fresh young leaves and berries. They consist of hederacoside, a saponic glycoside hederaginine and the hederasaponins A and B, which yield α and β hederin on hydrolysis. These are irritant to the stomach and intestine, causing vomiting and diarrhoea with nervous depression, narcosis and difficulty in breathing, which may require treatment, especially in young children. Dermatitis may result from contact with the foliage.

**Symptoms** Ivy can be irritant to human skin, causing reddening and dermatitis. Ingestion of seeds, berries or leaves causes irritation of the lining of the mouth, with blistering, swelling and pain. If a large quantity is eaten there will be nausea, vomiting and violent diarrhoea, with restless excitement and possibly difficulty in breathing. NPIS record eight cases with mild symptoms and uneventful recovery. Although the leaves are harmless when fed in small quantities to livestock, poisoning has been recorded in Jersey cows which had eaten large quantities of ivy; and in deer, sheep and dogs. Symptoms in cattle include staggering as if affected with milk fever, hyperaesthesia and bellowing as if in pain. Recovery is usually spontaneous, although their breath and milk will stink of ivy for several days.

**Treatment** Vomiting should be induced and demulcents given orally. Antihistamines will reduce the oropharyngeal oedema, and mild sedation may be required. Similar treatment may be needed for animals poisoned by ivy.

# · STRAWBERRY-TREE ·

***Arbutus unedo* L.**

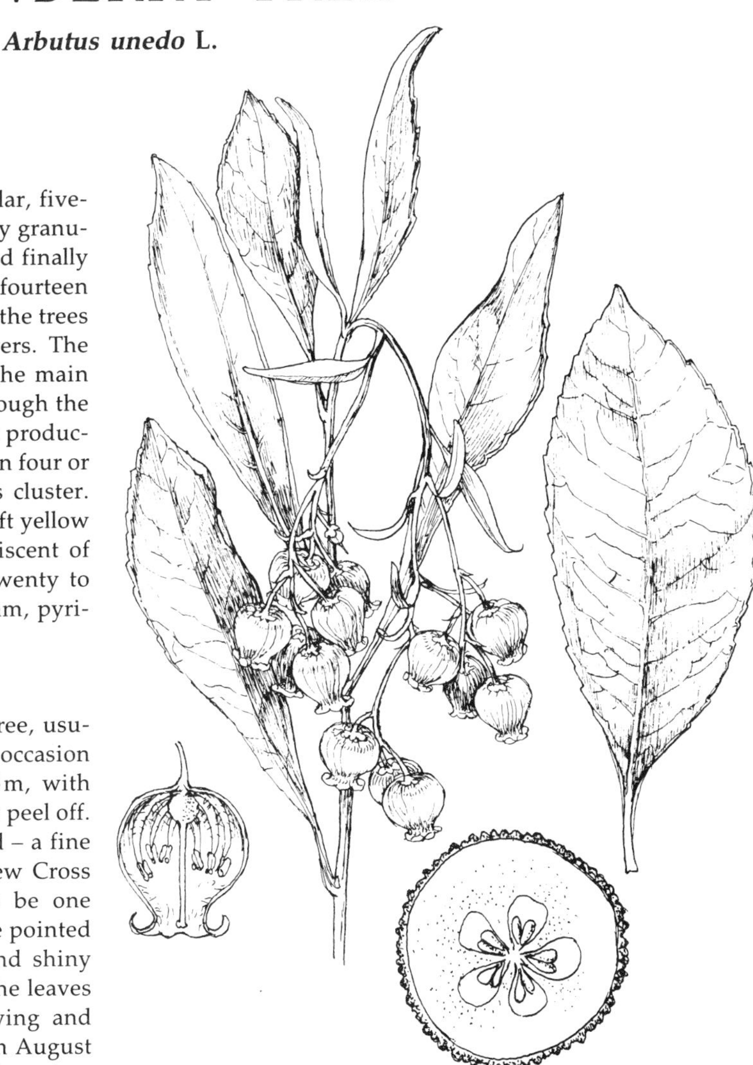

*Fruit*
The fruit of the strawberry-tree is a globular, five-celled succulent berry, covered with a warty granular skin. It ripens from yellow to scarlet and finally to deep crimson. The fruits take thirteen to fourteen months to mature, so they often appear on the trees at the same time as the next year's flowers. The earliest fruits ripen in mid-October, but the main crop is from November to December. Although the racemes contain forty to fifty flowers, fruit production is poor, and there are seldom more than four or five of the large fruits in each pendulous cluster. The berries are 20mm in diameter, with soft yellow flesh which has a pleasant flavour reminiscent of strawberries. A large fruit may contain twenty to forty seeds in the five cells, each 3.5 × 2mm, pyriform, finely striate and brown.

*Leaf and Flower*
Strawberry-tree is an evergreen bush or tree, usually a bushy shrub up to 3m tall but on occasion growing as a broad spreading tree to 15m, with rough, reddish-brown bark which tends to peel off. The trees are slow-growing and long-lived – a fine specimen 14m high in the grounds of New Cross Hospital in South London is known to be one hundred years old. The alternate leaves are pointed oval, with a serrated edge, dark green and shiny above and paler below. On mature trees the leaves last for twelve to fifteen months, yellowing and falling in August. Flowers are present from August to December, but maximum flowering takes place in October, the blossoms borne in small drooping terminal racemes of as many as fifty flowers. Each is 7-9mm long, with five minute triangular calyx lobes and a bell-shaped corolla which is pink or white. There are five ovate, spreading corolla lobes, the inside of the bell is white and hairy, and there are ten pink-flushed stamens with dilated, fleshy filaments. The stamens are hairy and there is a single style. The wind-pollinated flowers are visited by numerous bees and some butterflies.

*Ecology and Distribution*
Strawberry-tree is essentially a Mediterranean plant since it is sensitive to recurrent low temperatures, and seed production is poor following a hard winter. It is native in south west Ireland, growing in West Cork at Crookhaven and Glengariff, in South Kerry at Cloonee Lakes, Lough Guitane and Lough Currane, and especially on carboniferous limestone around the Lakes of Killarney. It occurs again on carboniferous limestone at the west end of

Lough Gill in Sligo. It is found on rocky lakeshores, cliffs and rocky slopes on the margins of oakwoods, disliking too much shade or a waterlogged soil. Elsewhere in the British Isles it is not infrequently grown in gardens and is naturalized on limestone cliffs in Wales – in Monmouthshire, Glamorganshire, and on Great Orme's Head in Caernarvonshire. It is naturalized and self-regenerating in a few Somerset woods, and occurs sparsely as an established alien in a number of southern counties east to Kent. In Europe it grows chiefly around the Mediterranean east to Cyprus, where a very few bushes still remain on the Akamas peninsula. Thrushes and pigeons eat the ripe fruits.

*History and Uses*
Strawberry-tree was cultivated by monks in western France and Ireland, the fruits being eaten or used for making a liqueur. It has been suggested that it is an introduced species, as some of the old sites are near monasteries and hermitages. It has been known in Ireland for at least three hundred years, and in 1586 young trees were sent from Killarney to England. Irish names for the strawberry-tree include caithne, cuince and caine apple.

**Strawberry-tree**

# · *BEARBERRY* ·

***Arctostaphylos uva-ursi* (L.) Sprengel**

*Fruit*
The fruit of bearberry is a berry-like drupe, borne in groups of two to five on very short pedicels. The fruit is bright red and shiny, about 8mm in diameter and 5mm tall, with the top flattened and dimpled-in, so that it resembles a tiny red doughnut. The long stylar remnant protrudes from the dimple at the top. The fruit is hard, with a dry, crumbly texture, and if cut across reveals five to seven cells, each containing a single pyrene 3 × 2.5mm, yellow, smooth and hard.

Bearberries are edible, ripening from August to October, but they are insipid and mealy. Fruiting is never heavy, occurring best at the edges of broken rocks or scree, where there is no competition.

*Leaf and Flower*
Bearberry is a strong-growing, perennial evergreen woody undershrub with trailing and branching stems usually prostrate up to 1.5m long. The leaves are small, entire, blunt ovate and thick, with well-marked veins and hairy edges. The trailing stems

are round, red and downy, and the leaves on them have twisted petioles, so that the leaf lies at 90° to the plane of its attachment.
Some of the leaves turn red in autumn, but the plant remains evergreen. Bearberry's main flowering period is from mid-May to mid-July, the pale-pink flowers borne in clusters of five to twelve. There are five sepals and a bell-shaped corolla with five spreading lobes. The flowers are strongly scented and are mainly pollinated by bees. In Glen Clova, Willis and Burkill observed the bearberry flowers are visited almost exclusively by various species of *Bombus*, especially *B.lapponicus*. The Arctic moth *Anarta cordigera* has also been recorded visiting flowers. A second flowering season seems to occur in late September when the fruits are ripe, but the flowers never develop into fruits.

*Ecology and Distribution*
Bearberry is an alpine, abundant in the lowest of the mountain communities and in the heather zone below the summits, flourishing best on the loose, stony edges of terraces. It is quick to colonize the broken edges of new road and railway cuttings through rocks, and likes bare, dry slopes where there is little competition. It is also part of the summit heath vegetation, growing with cowberry (*Vaccinium vitis-idaea*), bilberry (*V.myrtillus*), crowberry (*Empetrum nigrum*) and wavy hair-grass (*De-*

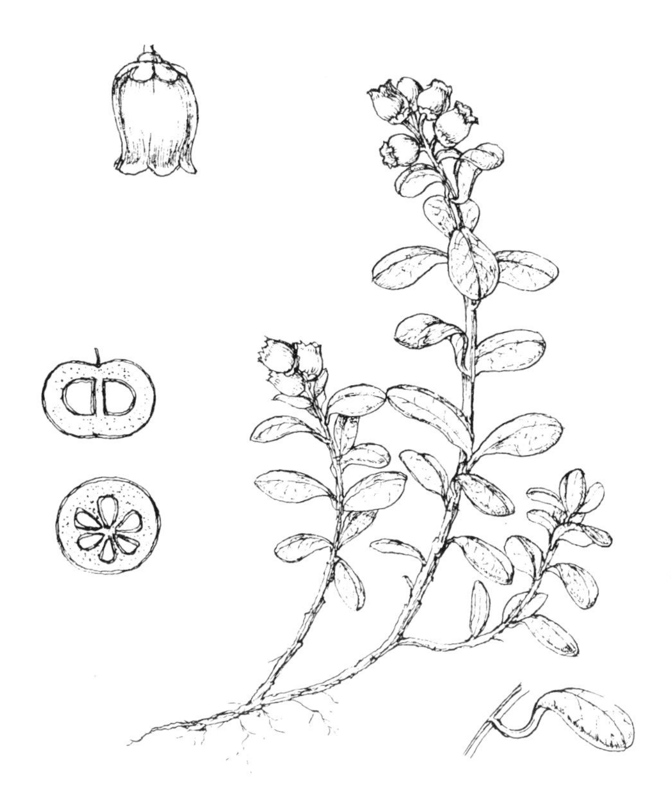

**Bearberry**

*schampsia flexuosa*), and it can be abundant where acid flushes drain from summit detritus. It is local but often abundant in the Highlands of Scotland, out to the Western Isles, and north to Orkney and Shetland where it is infrequent. It grows in a few isolated places in the Lake District, and south to High Peak in Derbyshire. It is absent from Wales, but in Ireland it flourishes on the Burren in Co Clare, on Slieve League in Donegal, and in the Kerry-Cork highlands. It has often been seen growing at sea level on Ardnamurchan Point in Argyll and at Culduie in Wester Ross, and this is also true of some of the sites in Donegal. In Europe it grows widely on heaths and mountains in most countries, except in the extreme south.

*History and Uses*
The dried leaves are powerfully astringent, and an infusion was used for the treatment of urethritis and cystitis. There are records of its use in the 13th Century by the Welsh 'Physicians of Myddfai'. The fruit is edible, but definitely not worth the effort!

# · ALPINE BEARBERRY ·

**_Arctostaphylos alpinus_ (L.) Sprengel**

*Fruit*
The fruit of alpine bearberry is a globose drupe 6-10mm in diameter, dark crimson when unripe, turning black and glossy when ripe in August and September. The colour gives the species its alternative name of black bearberry. The five calyx lobes remain beneath the ripening fruit, each lobe pointed oval and retaining a reddish-purple colour. The berries are fleshy, with reddish-purple juice and a sweet flavour like that of bilberry. They are five-celled, containing four or five oval, yellow pyrenes, each 3 × 2.5mm with a faintly pitted surface. Fruit production tends to be rather sparse.

*Leaf and Flower*
Alpine bearberry is a deciduous perennial, with prostrate, much branched, woody stems up to 60cm long. The new twigs are a beautiful red colour, and carry blunt, ovate leaves which are thick, slightly serrated and deeply marked and wrinkled with veins. The upper leaf surface is glossy and the lower leaf edges and petiole carry a few simple hairs. In autumn the leaves turn fiery crimson before they are shed; some wither and remain on the stems until the following spring. The brilliantly coloured leaves then make the plant easy to detect. The flowers appear from mid-May to mid-July, and are less conspicuous than those of bearberry, appearing to be self-pollinated. They are white, 4-5mm long and bell-shaped, carried in short racemes of two to four. The corolla has five ovate, spreading lobes.

*Ecology and Distribution*
Alpine bearberry is a rare plant of the high mountains of north west Scotland, preferring totally open, exposed areas where there is no competition from other herbage; here it is often found with dwarf azalea (*Loiseleuria procumbens*). It is not so much a plant of mountain summits as of dry, bare stony hill shoulders, nearly always on the 650m

**Alpine bearberry**

contour. It grows in a number of stations on or near Ben Nevis, notably on the edges of the summit plateau of Ben Chlianaig, in the east on Creag an Tarmachan south east of Strathspey, and it has been recorded from South Aberdeen, Argyll and mid-Perth, where it is at its most southerly locality on Sgiath Buidhe in Glen Lochay. Otherwise it is found entirely north of the Caledonian Canal, becoming increasingly common around Glen Affric and between Loch Maree and Loch Broome. As John Raven remarked, it is puzzlingly absent from the Cairngorms. It grows in a few sites in both Orkney and Shetland. In Europe, alpine bearberry can be found on heaths and stony slopes on mountains – in north Russia, Finland, Sweden and Norway, except in the extreme north, and on the mountains of central and south Europe, south to the Pyrenees, central Apennines and northern Albania.

# · CRANBERRY ·

***Vaccinium oxycoccos* L.**

*Fruit*

Cranberries are borne on reddish-brown, wiry 25 mm pedicels, so fine that when the fruits are fully developed and ripe, from late August onwards, they fall into the moss cushions in which they grow and are difficult to see. The ripe berries are 6-10 mm in diameter, either globular or pear-shaped, with a blunt apex from which the curled remnant of the stigma protrudes. The berry is pinkish-buff, with a thin translucent skin covered in minute red and brown, irregularly shaped dots. The texture is fairly hard, with translucent flesh and a slightly acid flavour, like that of bilberry but not as strong. The berries are never juicy. Their shape is variable. In a

bog near Perth, most of the fruits were of moderate size, egg-shaped and mottled, while a few were larger and pear-shaped with the upper surface red and the lower yellow-green. Although the berries ripen from late July onwards they may remain on the plants in good condition until March of the following spring, beginning to wither and collapse in May. The berry is divided into four loculi, each containing three or four pointed, oval seeds about 2mm long, brown with a honeycomb surface.

*Leaf and Flower*
Cranberry is a slender perennial with fine wire-like stems which creep and root. The evergreen leaves are very small, 4-8mm long, alternate and blunt oval, with nearly parallel sides. The upper surface of the leaf is dark green with the margins slightly rolled over. The under surface is white, marked with tiny dark dots. The flowers appear from mid-June to mid-August, one to four in a terminal stalked inflorescence, each on a long erect pedicel which is usually hairy and bears two tiny pink bracts at or below the middle. The calyx is campanulate, and the four rosy-pink petals, 5-6mm long, are almost free, folding back like a miniature turk's-cap lily. In the centre of the flower there is a projecting bunch of eight yellow stamens, with a longer stigma in the middle.

**Cranberry**

*Ecology and Distribution*
Cranberry grows in acid sphagnum bogs throughout the wetter parts of Britain, but it is never common. In times past it had a wider distribution in southern England, becoming extinct in Suffolk by 1810 and Kent by 1899. It still hangs on in three sites in Sussex, and in Somerset now only occurs in blanket bogs on Exmoor between the Devon border and Dunkery Hill, although it may still exist on Mendip where it was last recorded at Blackdown in 1957. The main centre of distribution now is in southern Scotland, where it is quite abundant on some of the islands in Loch Lomond and on the Campsie Hills near Glasgow. It also grows in the

Borders and fairly extensively in Wales, especially in north and central Wales. There is a recent record (1976) in north Lincolnshire at Greetwell-in-Manton, and it still grows at Wybunbury Moss in Cheshire, but in the far north of Scotland it is a rarity. In 1983 it was found growing in a raised bog at Strathbogie in North Aberdeen. Most of the records for Ireland are for the south east. In north and central Europe it is a plant of peat bogs, usually in the wetter areas, extending locally to south central France, north Italy and south east Russia.

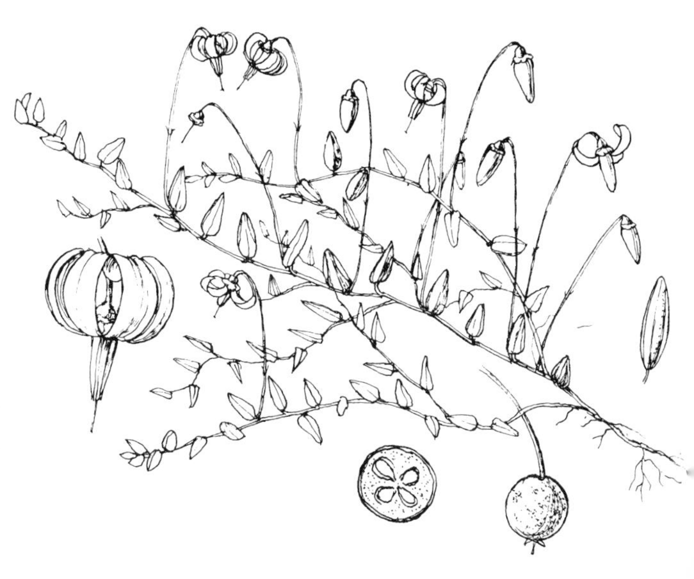

*History and Uses*
The berries of the British species are virtually inedible raw but make a good sauce which will keep well until winter time. The cranberries used commercially for sauce come from a North American species.

### *Vaccinium macrocarpon* Aiton

This cranberry comes from the east of North America and has been widely cultivated and locally naturalized in the British Isles, as well as in Germany, Switzerland and the Netherlands. It is more robust and taller than *V.oxycoccos*, with broad flat leaves nearly twice as long and flowers which are not carried in terminal spikes. The fruit is much larger, 10-20 mm wide, and redder. It was first mentioned as a naturalized plant in Britain by J.E. Lousley in 1936; by 1966 A.H. Sommerville recorded it as well-established on Arran in the Firth of Clyde. At Ashley Heath near Ringwood in Dorset it grows in a bog community with *Erica tetralix*, *Myrica*, *Sphagnum* and *Polytrichum commune*. However, it seems to tolerate drier conditions than *V.oxycoccos*, growing on Kingsley Common in Hampshire among heather and moor-grass and at Beaulieu Heath, but fruiting is sporadic. It has recently been recorded growing in a *Molinia* bog near Fishguard, Pembrokeshire, where it may have existed for at least twenty years.

# · SMALL CRANBERRY ·

**_Vaccinium microcarpum_ (Turczaninow ex Ruprecht) Schmalhausen**

*Fruit*
Small cranberry is, as the name suggests, smaller in all its parts than cranberry. The fuit is 5-9 mm wide, flattened globular, pear-shaped or sometimes lemon-shaped, red-tinged and lacking the spots of colour on the skin typical of cranberry. Fruits found in late August between Grantown and Forres in Grampian were 9 mm wide by 7 mm tall and deep crimson with a faint mottling of the skin. The flesh was crisp with a sour bilberry flavour and the fruit was divided into five loculi, each containing two pale yellow, pointed, oval seeds 1.5 mm long.

*Leaf and Flower*
Small cranberry is a prostrate, creeping evergreen with fine stems which rarely exceed 30 cm in length. The alternate leaves are smaller than those of cranberry and have a rather triangular outline, with the margins slightly rolled. The flowers open in July, and are similar in shape to those of cranberry, but

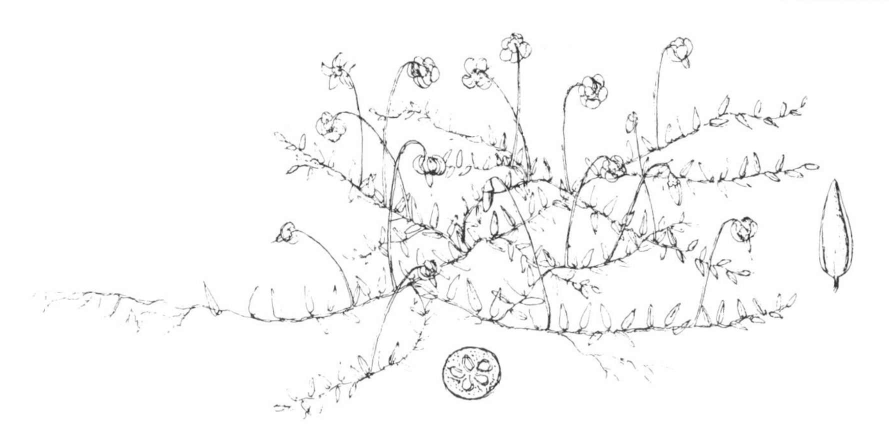

are smaller and darker pink, with hairless pedicels. This characteristic has been used to differentiate small cranberry, but around Loch Linnhe and near Bellanoch, Knapdale the separation of the two species by the hairiness of the pedicels has proved unreliable. All gradations exist between plants with hairy pedicels and those with entirely smooth pedicels. The size and shape of the leaves is equally unreliable as a means of identification. Chromosome studies show that the two cranberries are truly different species, cranberry being tetraploid and small cranberry diploid.

*Ecology and Distribution*
Small cranberry occurs in acid peat bogs with sphagnum moss and *Calluna*, growing in places rather drier than those usually favoured by cranberry. It is mainly found in the Highlands of north east Scotland, in Perth, where it grows on Sgiath Buidhe, and Inverness, particularly the area between Aviemore and Grantown-on-Spey, although it is also found in Argyllshire, on Ben Tee near Fort Augustus, Cairnbrallen in North Aberdeen, and north to Caithness. It is an uncommon plant and, like cranberry, is easily overlooked. It occurs widely in north and north central Europe as far south as the Alps, the Carpathians and north west Ukraine.

**Small cranberry**

# · COWBERRY ·

**_Vaccinium vitis-idaea_ L.**

*Fruit*

The fruit of cowberry is a shiny red, round berry 5-10mm in diameter. The berries hang on short pedicels 2-3mm long underneath the foliage, with two small cup-shaped bracts at the base of the berry. The four or five calyx lobes are shed well before the fruit is ripe in September and October. The apex of the berry comes to a blunt point and bears a prominent collar of the old floral attachments. The berry has four or five cells, and contains five to eight oval seeds about 1.5mm long, yellowish and faintly pitted. The berries are crisp in texture and contain an acid-tasting, non-staining juice.

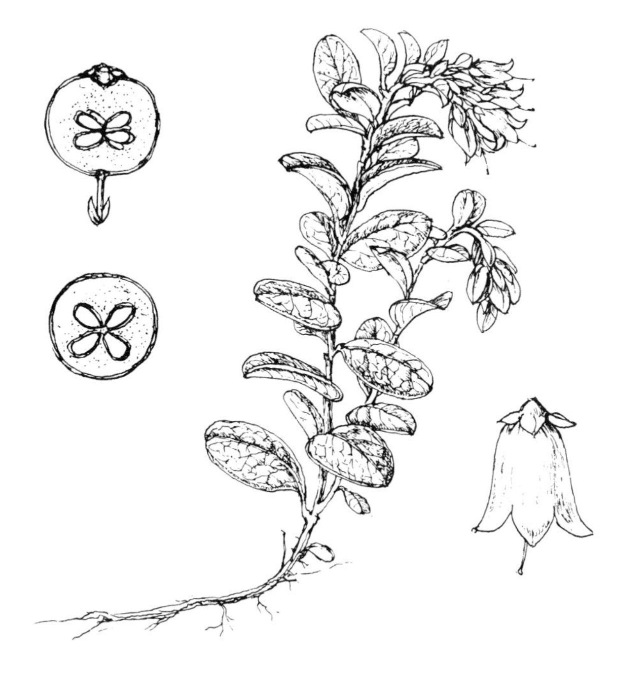

*Leaf and Flower*

Cowberry is a small evergreen shrub up to 30cm tall, with subterranean creeping rhizomes and erect or prostrate arching stems with rounded twigs. The alternate, blunt oval leaves are dark green, leathery and thick, the upper surface being glossy with prominent veins. The leaf edges are entire, with margins slightly rolled down, and the underside is pale, marked with dark glands. Immature leaves tend to be smaller and flatter, with less prominent veins and fewer glands on the underside. Cowberry flowers from late May to mid-June, with white, bell-shaped flowers 6-8mm long, tinged with pink, in short drooping terminal racemes of four to twelve. The calyx lobes soon fall off, and the triangular corolla lobes spread out. There are eight to ten stamens. In some years there is a second flowering period in September and October, the flowers being smaller than those of the main season, and rarely setting fruit.

*Ecology and Distribution*

Cowberry is one of the major northern montane dwarf shrubs, growing in a wide range of upland and mountain communities, from largely acid soils to pure peat bog. It seldom occurs below 200m and extends up to 1000m. It is one of the first flowering plants to colonize bare scree, with bilberry, crowberry and *Lycopodium selago*, and is a typical member of the flora of summit heath and bilberry moor. It flourishes in mossy pinewoods such as those of Rothiemurchus, where it grows tall and fruits well. In unshaded places it is characteristically found on rock ledges and screes. Where it occurs on summit heaths around 800m, as in Glen Roy, it can be dwarf and totally prostrate. It occurs widely in the British Isles, especially in the north of Scotland, where it has its maximum abundance in the pinewoods of the Highlands. It is found also in the Hebrides, Orkney and Shetland, and in Wales it grows mostly in the mountains of the north. In Ireland it is rare and local, being absent from the south west and much of the central plateau. It grows well on the north face of Slieve League by Donegal Bay. South from Scotland it is abundant on the moors in the south east Pennines and in many

**Cowberry**

places in Cumbria. It grows as far south as the Charnwood Forest north west of Leicester and Cannock Chase in Staffordshire. It is very rare in south west England, with one small area at the north end of the Quantocks in Somerset, where it was rediscovered in 1958, another site on Dartmoor where it certainly fruited in 1984. It was also recorded in 1971 in a woodland ride near Darwell in East Sussex. In north and central Europe it occurs widely on moors, heaths, coniferous forests and sub-alpine pastures south to the mountains of the Apennines, Albania and Bulgaria. Cowberry is slow to mature, and plants do not have many flowers until they are over five years old. Bees and wasps act as pollinators, and self-pollination also occurs. The fertile hybrid with bilberry, *V.myrtillus* x *vitis-idaea*, has been recorded from Yorkshire, the north Derbyshire moors and Cannock Chase in Staffordshire, where it was first reported in 1870.

*History and Uses*

The scientific name *vitis-idaea* means 'vine from Mount Ida', which is in Turkey, and where incidentally the species does not grow! Cowberry berries are scarcely edible when raw, and have a rather bitter taste. They make an excellent jelly, but apple must be added to provide the necessary pectin. The berries are eaten by grouse, ptarmigan, partridge and pheasant. Launert lists the ripe berries, in either the fresh or dried state, as a remedy for diarrhoea. The dried leaves have astringent properties and have been used to treat diarrhoea, and also as a remedy in rheumatic conditions.

# · *BOG BILBERRY* ·

**_Vaccinium uliginosum_ L.**

*Fruit*

The berries of bog bilberry are borne singly, or less often in groups of two or three, on pedicels 3-4 mm long. They are more egg-shaped than those of bilberry, averaging 10 × 8.5 mm, blue black with a waxy grey-blue sheen to the skin. The apex of the ripe berry frequently bears a long projecting stigma remnant and lacks the large, flat crater so characteristic of the truncated berries of bilberry. The depression is shallow, scarcely 4 mm across. The berry is divided into five loculi and the numerous pointed oval seeds measure 1.5 × 1 mm, pale brown, with a finely striated surface. The fruit is ripe from late July to September, and has a sweet, mild taste, lacking the pungency of bilberry. The flesh of the berry is translucent until fully ripe,

**Bog bilberry**

when it takes up a reddish-purple tinge from the skin, the resulting juice staining a deep bluish purple.

*Leaf and Flower*
Bog bilberry is a rather small, much branched perennial 15-45cm high, with straggling stems growing from creeping roots to form a low, broad bush. The twigs are round and shining yellow-brown, quite unlike those of bilberry. The leaves are ovate, entire edged and deciduous, broader in the distal third, and net veined. They have a distinctive blue sheen, which is helpful in rendering these otherwise inconspicuous plants visible, and in autumn the leaves turn an attractive mauve.

Bog bilberry flowers from mid-May to mid-June, but in the British Isles it is a shy flowerer – unlike the rest of northern Europe – and fruit is rarely produced. The flowers are 4-6mm long, borne in groups of one to four in the leaf axils, with short calyx lobes and a pale pink, ovoid corolla with very short, triangular, reflexed lobes.

*Ecology and Distribution*
Bog bilberry prefers the edges of broken blanket bog where the ground is damp, but there is a bank-like slope. In summit peat bogs it occurs with *Eriophorum vaginatum*, cloudberry, crowberry and *Salix repens*. The flowers are bee pollinated. In Scotland bog bilberry occurs on most mountains above 900m, particularly where a plateau edge exists with an accompanying wreath of rather prolonged snow-lie. In Glencoe and on Clova it grows on or near calcareous rock ledges at a considerably lower elevation. It is a local plant, though in places it can be abundant. It is rarely found in south Scotland with the exception of sites near Selkirk, and more recently (1970) in Dumfriesshire. It is recorded in a few sites in Cumbria and at Baron House Bog, west of Crag Lough in Northumberland. It does not grow in Wales and Ireland. In

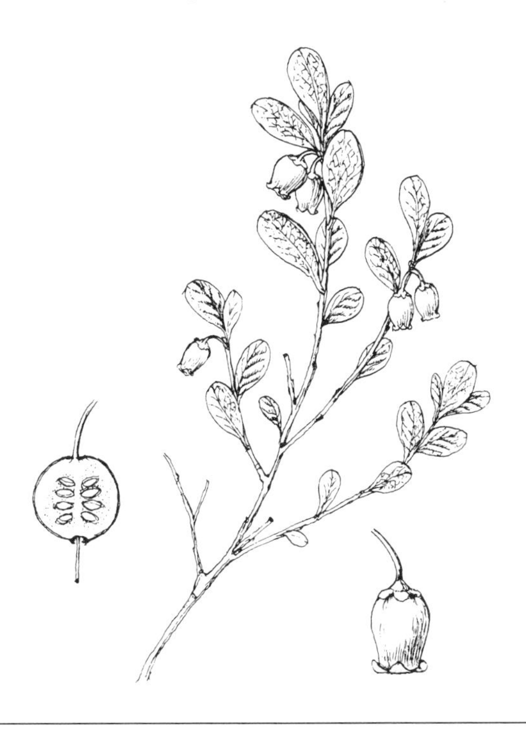

Europe it grows on moors and heaths, in coniferous woods, subalpine pastures and tundra in north and central regions, south to the Sierra Nevada, north Apennines, Albania and Bulgaria.

*History and Uses*

Like so many of the British montane plants, bog bilberry fruits very poorly in Great Britain, yet in Norway and Sweden it bears abundant fruit. The berries are particularly succulent, although less sweet and tasty than those of bilberry. They can be eaten raw, stewed or used in making preserves. The British botanist will be hard pressed to find ripe fruit, while his Swedish counterpart can pick enough to make jam! The Arctic form is diploid (2n = 24), while the form common in north Europe, and possibly the form found in Britain, is tetraploid (2n = 48). The polyploid form shows greater vigour and it spreads vegetatively with ease.

# · *BILBERRY* ·

**_Vaccinium myrtillus_ L.**

*Fruit*

Bilberries are borne singly or in small groups in the leaf axils on tough green pedicels 7mm long, tinged with brown where they join the base of the globular berry, 3-10mm wide and slightly less in depth. The fruit is purple-black with a blue bloom on the smooth skin, and the top of the berry bears a shallow crater 2-4mm wide which initially is ringed by the shrivelled remnants of the flower. Bilberries are sweet and juicy when ripe from July to September, with reddish translucent flesh and juice which stains a deep, bluish purple. The berry is usually five-celled, each cell containing numerous small, oval, yellow seeds, 1mm long, and marked with longitudinal striations. Most berries contain approximately twenty seeds.

*Leaf and Flower*

Bilberry is a much-branched deciduous shrub, 10-60m high, with erect branches. The stems are green and three-angled, bearing light green, ovate leaves, rather thin and with serrated margins. The leaves turn yellow-red and then brown in autumn. The stems carry two types of bud – one enclosing two bracts and a single flower, and the other containing a single lateral shoot with a flower in its axil. Flowers appear from late April to the end of June, usually solitary and about 6mm long, with five small, sinuate calyx lobes and a bell-shaped pink corolla tinged with green, bearing five triangular, recurved lobes. There are ten stamens.

*Ecology and Distribution*

Bilberry is abundant on heaths, moors and oak, birch and pinewoods on acid soils throughout north west Britain at altitudes up to 1200m. In the south of England it grows locally on the Lower Greensand, Bagshot Sand and in the Weald, appearing very locally on heaths in west Kent, on the heaths and acid woodlands of the more northerly parts of Sussex and then abundantly in the New Forest and on the heaths of Dorset. In Somerset it is abundant on Exmoor and the Blackdown Hills. It is absent from a large area of south central and eastern England which corresponds to the major areas of chalk and limestone. In Wales, the Pennines and Cumbria it is common on gritstone edges and screes, and from here northwards it is abundant. In Ireland it is common, especially in the mountainous areas. The dwarf shrub zone, composed of bilberry and crowberry, is one of the main features on mountains above the heather zone and below the open summits, and it may form isolated island communities in sheltered, well-drained places on otherwise exposed bare tops, especially where rocks break the wind. They are a characteristic feature of the Cairngorms at about 1000m, although they are absent from patches where the snow lies late in the season. This zonation is not as clear on the Welsh mountains as it is in Scotland, except in a few places such as Y Garn above Cwm Idwal.

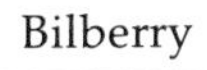

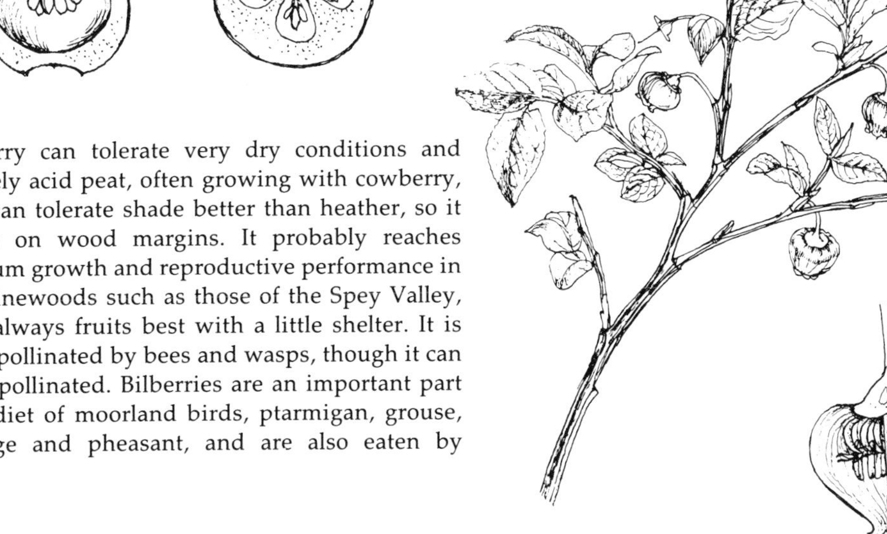

Bilberry can tolerate very dry conditions and extremely acid peat, often growing with cowberry, and it can tolerate shade better than heather, so it persists on wood margins. It probably reaches maximum growth and reproductive performance in open pinewoods such as those of the Spey Valley, and it always fruits best with a little shelter. It is chiefly pollinated by bees and wasps, though it can be self-pollinated. Bilberries are an important part of the diet of moorland birds, ptarmigan, grouse, partridge and pheasant, and are also eaten by

**Bilberry**

mistle thrushes and song thrushes. Blackbirds have been recorded eating them, particularly in Wales. In Europe bilberry is widespread on heaths, moors and acid woodland, but in the south it is restricted to the mountains.

*History and Uses*
As befits a fruit which is both widespread and excellent to eat, bilberry enjoys a multitude of common names. The Scots name of Blaeberry derives from the old Norse 'blaa', meaning dark blue, while the southern name of Whortleberry may be a corruption of Myrtleberry. Bilberry is also called Whinberry and Huckleberry, and the fruit is sometimes called a hurt or whort, hence the name Hurtwood Hill, near Leith Hill in Surrey. Bilberries are excellent eaten raw and make splendid pies, puddings, jellies and jams. They are rich in Vitamin C. Gent (1681) wrote: 'Black bilberries are good for hot Agues, and to cool the heat of the Liver and Stomach, and do bind the Belly'. Launert recommends two tablespoonsful of dried berries to be taken once daily for enteritis, or the same quantity of the fresh fruit for constipation. The dried fruits have been used as a diuretic and to treat degenerative retinal conditions. Decoctions made from the leaves are unsafe, as they contain hydroquinone. In Germany bilberry juice was used in the past for dying cloth.

## *Vaccinium corymbosum* L.

This species, known both as blueberry and swamp blueberry, was first grown in Britain in 1765 when it was introduced from eastern North America. In 1980 a naturalized colony was discovered by R.P. Bowman on Ashley Heath near Ringwood, Dorset, on damp sandy soil. It is a low, bushy, deciduous shrub growing to 1.2m, (in America it can reach 4m), with leaves resembling buckthorn, colouring to shades of red in autumn. The cylindrical, pale pink flowers are produced in May in few-flowered clusters near the leafless ends of the previous season's twigs. The fruit is 12 × 10mm, blue-black with a marked bloom on the surface, ripening in July and August.

# · CROWBERRY ·

**_Empetrum nigrum_ L.**

*Fruit*
The fruit of crowberry is a slightly flattened, irregular globular drupe, 5-8mm in diameter and 4-5mm tall, with a dimpled apex. The skin of the berry is green when first formed, changing colour through pink and purple to glossy black when ripe in early July. Ripe fruits remain long on the plant and may persist over winter. They are borne singly or in pairs on short pedicels. The fruit is solid in texture, with purple juice, and contains two to nine stones, 1-2mm long, triquetrous with a rough surface finely netted in dark red. The fruits are scarcely edible. In the absence of foliage it is difficult to differentiate this species from mountain crowberry, but a careful examination of the base of the fruit will reveal the shrivelled remnants of two whorls of small perianth segments, the outer whorl broad oval in outline.

*Leaf and Flower*
Crowberry is a low, heath-like shrub, with prostrate, rooting stems spreading to form a dense mass of foliage. It can grow up to 50cm high, but plants in mountain summit heath will be dwarf. The young twigs are usually reddish. The leaves are alternate, in a close-set spiral, each 5-7mm long and 1-2mm wide, with the surface folded over into a tube with a ventral slit. The leaves are three to four times as long as they are broad, with parallel sides and moderately long internodes, though seaside plants exposed to salt wind may be close-growing and resemble mountain crowberry.

Crowberry is normally dioecious, the male and female flowers appearing on separate plants in the axils of the upper leaves from late March to May, depending on the altitude. The male flowers are reddish, 1-2mm long, with six perianth segments and three projecting stamens with red anthers. Male plants bear numerous flowers, usually massed towards the top of the stem. The female flowers are small and inconspicuous, up to six on a stem, usually solitary, with dark violet, sessile stigmas. The flowers are wind-pollinated. Plants with hermaphrodite flowers have been recorded rarely. The normal form of crowberry is diploid (2n = 26).

*Ecology and Distribution*
Crowberry is common on the wetter and hilly moorlands of the north and west of the British Isles, at altitudes from sea level up to 800m, where it overlaps in distribution with mountain crowberry. With bilberry it is the main constituent of the dwarf shrub community above the heather zone on mountains, but below the summits. It is tolerant of some snow cover – but only where this disappears quickly in spring, since it flowers early in the season. It grows on dry heath with *Calluna*, on wet heaths, acid grassland and in open pine and birch forest with bilberry and *Calluna*, and is an impor-

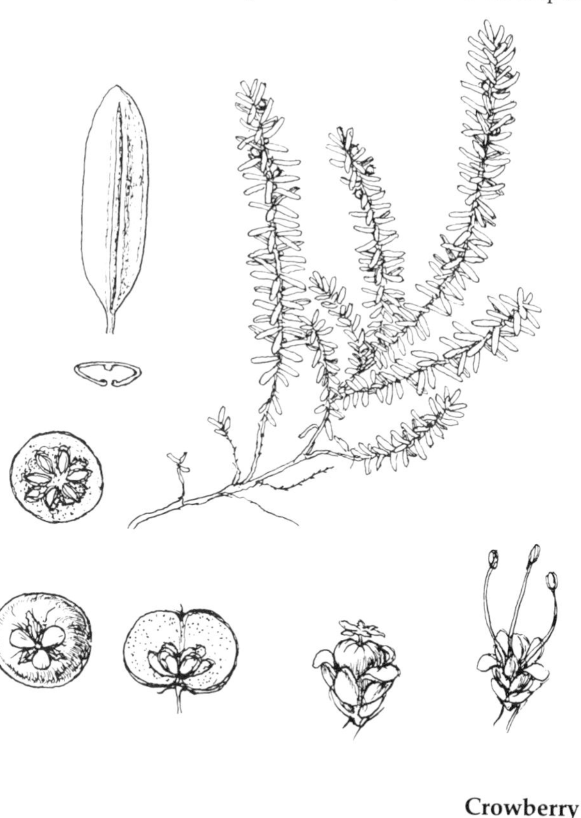

**Crowberry**

tant colonizer of leached acid soils. In the South of England it occurs on a few high moors on Dartmoor, Exmoor and on Haddon Hill, and in the east in two isolated localities on the coast of Norfolk. It is common on moorland in Central and North Wales, Northern England, Scotland, the Hebrides, Orkney and Shetland. Although present throughout Ireland it is more plentiful in the north. Although generally a plant of acid soils it grows on limestone pavement in the Burren in west Ireland, and on calcareous shell sand in Sutherland. In Europe it grows widely on heaths, moors, stony mountain slopes and in coniferous forests, but is absent from the extreme north and from most of the Alps.

*History and Uses*

Crowberry was abundant on acid soils in the Late Glacial flora all over the British Isles, but especially in the Irish Allerød sites. The species found in deposits is usually the diploid *E.nigrum*. The fruits are scarcely edible, but can be used for making jelly. They are not touched by deer or sheep, but the leaves are eaten by grouse and ptarmigan, and the fruits by many animals and birds including foxes, geese, crows, plovers, thrushes, waders, gulls, terns and skuas. The seeds will pass undigested through the alimentary tract of both animals and birds, and can often be seen in the 'cigarette-end' droppings of grouse and ptarmigan.

# · *MOUNTAIN CROWBERRY* ·

***Empetrum hermaphroditum*** **Hagerup**

There is ample evidence to separate the two crowberries by their distribution, morphology and chromosome number, mountain crowberry being tetraploid (2n = 52).

*Fruit*

The fruits of mountain crowberry are flattened, black globular drupes like those of crowberry, and of a similar size, 5-8mm in diameter and 5mm tall. They are ripe from July onwards.The fruit is rather solid in texture and contains two to nine stones, 1-2mm long, rough and triquetrous. It is possible to differentiate the fruits from those of crowberry by examining the perianth remnants at the base of the fruit. Since this species is hermaphrodite there should be some stamen remnants between the berry base and the perianth, but by the time the fruit is ripe they have nearly always fallen off. However, the perianth remnants are smaller than those of crowberry, the outer whorl of three being small and narrow, so that the two whorls together have a markedly triangular outline. The berries have a sweet flavour and when they are fully ripe the purple-staining juice is pleasant and refreshing.

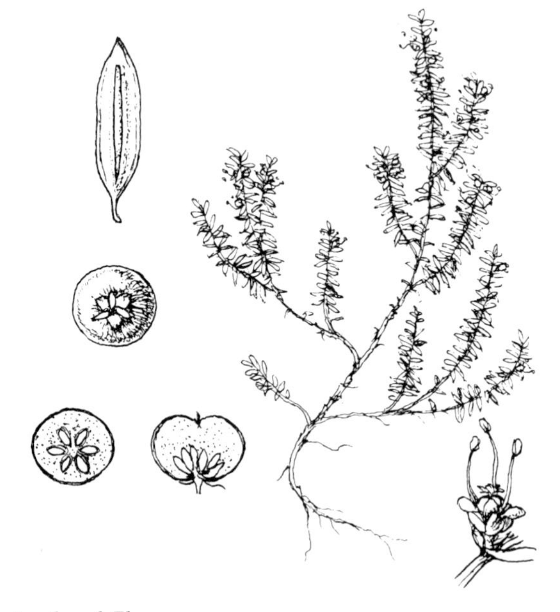

*Leaf and Flower*

Mountain crowberry is a low, heath-like shrub, forming dense masses up to 50cm high, with denser, yellower, more upright foliage than crowberry, and with a distinctive feel when walked upon which can only be described as crisp! The young

stems may be green or reddish, which is little help in determining the species,but the leaf internodes are short, so that the plant looks denser and bushier than crowberry. The leaves measure 3.5-6mm long and 1.5mm wide with the margins rolled over to form a tube, but they are two to three times as long as they are broad, and they taper markedly from the proximal third to the tip.

The flowers appear from late April to early June, 1-2mm long, massed in the leaf axils towards the top of the stem. They are hermaphrodite, with six reddish or purple, rounded perianth segments, projecting stamens and a short style.

*Ecology and Distribution*

Mountain crowberry is a montane plant, growing at elevations of 700m to 1200m. It is widespread and common on the mountains of Scotland, especially north of the line between Glasgow and Perth. In most high altitude sites it displaces *E.nigrum*. It is characteristic of the summit heath vegetation of the

**Mountain crowberry**

Cairngorms, growing with cowberry, bilberry and bearberry, and it is often so dwarf that the ripe fruits appear scattered on the ground like shiny black sheep droppings. It also grows in high altitude bogs with cloudberry and bog bilberry. It is found in two sites in Westmorland and one in Cumberland. In North Wales it occurs on the Glyders, on Snowdon, Craig-yr-Ysfa and Pen-yr-helgi-ddu and on the north face of Cadair Idris. It is uncommon in the Hebrides and Shetland and absent from Ireland. In Europe it grows in the north down to lat 60°N, and on some of the high mountains further south.

*Uses*

Mountain crowberry is eaten by a wide variety of animals and birds, especially by ptarmigan.

# · WILD PRIVET ·

***Ligustrum vulgare* L.**

POISONOUS

*Fruit*
The fruit of wild privet is a shiny black, rounded berry 6-8mm in diameter, slightly irregular in shape and carried in dense erect clusters on the ends of the twigs. It has a leathery skin beneath which is a thin, purple-staining layer of flesh which turns powdery when handled. The berry is two-celled, sometimes developing more than one seed in each cell, but the normal number of seeds is one to three per berry. The seeds are large, 5 × 4mm, shiny and black with a pitted surface, and with flat facets where they abut. They are often of very unequal size. The fruits are ripe from late July onwards, and in mild winters they will remain on the bushes until the end of February of the next year.

*Leaf and Flower*
Wild privet is a spreading, half-evergreen bush, 1-5m high. The summer leaves are opposite, pointed elliptical, with a smooth margin. The over-wintering leaves tend to be broader near the apex and often show bronzing. The flowering season is from mid-June to late August, the white flowers produced in masses of panicles at the tips of the stems. Each flower is 3-4mm long, with a tubular four-toothed calyx and a tubular corolla with four petal-like lobes as long as the tube. The flowers have a sickly-sweet scent and are insect-pollinated. The evergreen Japanese privet *L. ovalifolium* Hasskarl is widely grown as a hedging plant. It has larger, broad oval leaves, and flowers in which the corolla tube is longer than the petal-like lobes.

*Ecology and Distribution*
Wild privet is native as a plant of scrub on chalk and limestone soils, but elsewhere it is uncommon. It sometimes grows in fen carr. It is widely planted and may not be native in Scotland, where it is

**Wild privet**

absent from the montane areas, and from Skye and Orkney. In Ireland it is probably only native in Waterford, North Tipperary, south east Galway and Dublin. In Europe it is a plant of wood margins and scrub, preferring calcareous soils, growing in south, west and central Europe north to 59'N and in a small area of south east Norway and south west Sweden.

*History and Uses*

Wild privet is mainly useful as a hedging plant. Gent wrote: 'Prim or Privet, the Leaves and Roots of it are binding and is good to wash sore Mouths, to cool Inflammations, and to dry up Fluxes, and is good for Ulcers in the Mouth and Throat, and all swellings and Impostumes, and is good against all Fluxes of the Belly and stomach, and bloody Flux.' Deakin noted that the berries were eaten by pheasants and partridges, but blackbirds are the species most frequently recorded eating them.

*Toxicology*

The berries and all aerial parts of the plant are poisonous, containing the glycoside ligustrin which has an irritant action upon the stomach, intestines and kidneys. The bruised leaves and crushed berries can cause acute contact dermatitis in humans.

**Symptoms** Ingestion of privet causes nausea, vomiting and profuse diarrhoea, with colicky abdominal pain because of intense gastro-intestinal inflammation. Fluid loss and dehydration can be a severe problem. Poisoning, though rare in humans, can be fatal in children. In animals it is uncommon, though fatal poisoning has been recorded in sheep and horses, which exhibited colic, ataxia and paralysis of the hind quarters, and died four to forty-eight hours after eating hedge-trimmings. Poisoning of cattle has been recorded in New Zealand. NPIS list eight human cases, all exhibiting mild symptoms, and recovering without sequelae.

**Treatment** Vomiting should be induced, if it has not already occurred, and activated charcoal should be administered orally. Fluid replacement therapy may well be indicated, as well as supportive treatment as necessary. In cases of dermatitis a soothing lotion should be applied to affected areas.

# · DUKE OF ARGYLL'S TEA-PLANT ·

**_Lycium barbarum_ L. (inc. _L.halimifolium_ Miller)**

POISONOUS

*Fruit*
The fruit of Duke of Argyll's tea-plant is a soft orange berry, bean-shaped, 8 × 7mm, with a distinct groove on either side and across the apex of the fruit. The berry is borne on a long pedicel and sits in a cup formed by the two green calyx lobes. Its flesh is thick and orange-coloured and has a sweetish taste. The berry is divided into two cells, and normally only one seed matures in each cell. The seeds are blunt, disc-shaped, 3 × 2.5mm, with a netted brown surface. The berries are usually ripe in October. In most seasons very few are produced.

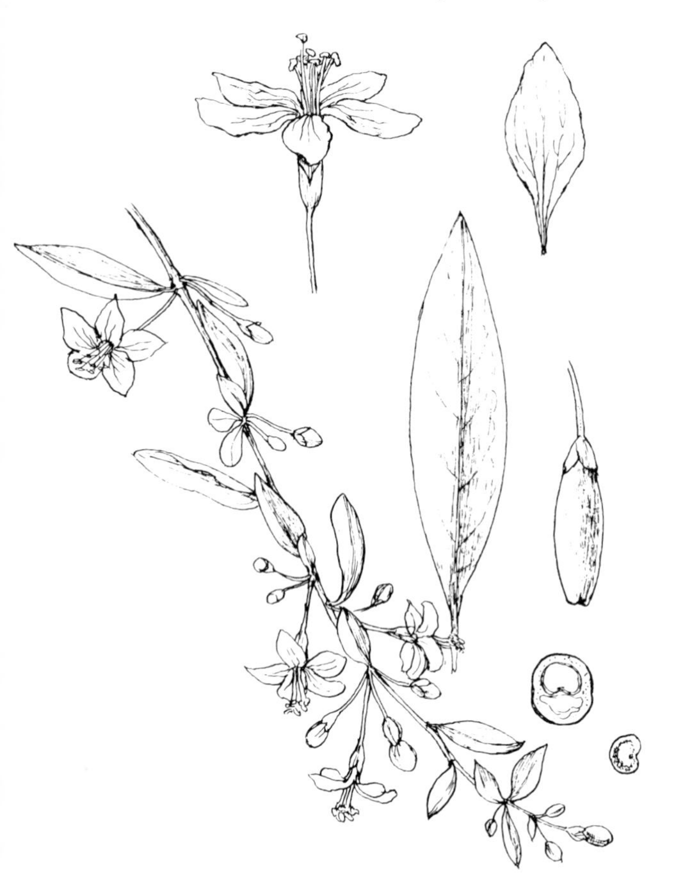

*Leaf and Flower*
Duke of Argyll's tea-plant is a small deciduous shrub up to 2.5m high with arching stems which may bear a few spines, and rather greyish-looking foliage. The leaves are alternate, lanceolate, widest at the middle, with a leathery texture and dull grey-green surface. The species shows considerable variation in growth characteristics, plants in sheltered sites on good soil tending to have larger leaves and stems with fewer, smaller spines. The flowers are borne solitary or in small groups of up to three in the leaf axils, from mid-June to October, so that flowers and fruit are present at the same time. They are typical Solanaceous flowers, with a tubular calyx, usually two lobed, a mauve or pinkish-purple narrow, tubular corolla with five spreading lobes and projecting stamens and stigma.

*Ecology and Distribution*
Duke of Argyll's tea-plant is introduced and widely distributed in England and Wales, less commonly in Scotland where it grows as far north as Durness in Sutherland. In Ireland it is uncommon and found mainly on the east coast. It has been extensively cultivated and naturalized and flourishes especially well in sandy areas near the sea in southern England. Most of the records come from coastal sites, particularly in north Kent, the coast of Suffolk, and in warm and sheltered spots around the coasts of Wales. Many of the sites are near abandoned cottage gardens. In Europe it is cultivated for hedges and is widely naturalized.

## *Lycium chinense* Miller

This species was introduced from China in 1700. It has larger, brighter-green leaves which are wider below the middle, and is less spiny. The flowers are produced later than those of *L. barbarum*, from August to late October, though there is some overlap in the flowering period. The corolla lobes equal the tube in length and the calyx is much shorter. *L.chinense* will be found in similar places to *L.barbarum*, but appears to be less common; for example it has been recorded in a few sheltered sites in Iona off the south west tip of the island of Mull, Argyll.

*History and Uses*
*Lycium barbarum* was introduced into the British Isles from the Near East in 1696 under the name of *L.halimifolium*. The common name is supposed to have arisen following a nurseryman's mistake when supplying plants to a former Duke of Argyll who had ordered Tea-plants. Hardin and Arena state that the fruits are edible raw or cooked, but it is probably wise to avoid them in spite of their pleasant flavour.

*Toxicology*
There is some confusion over the possible toxic properties of this species. Some authorities state that the berries contain an atropine-like substance, tropeine, as well as hydrocyanic acid and betaine; others maintain that the fruits are edible but that the leaves contain the toxins.
**Symptoms** These alkaloids have parasympatholytic properties, so that likely symptoms of poisoning would include visual disturbance, thirst and a dry mouth, reduced gut motility, increased urination and a dry, flushed skin. No cases of human or animal poisoning have been reported, but, in view of the confusing reports, the fruits of this species should be avoided.
**Treatment** Vomiting should be induced, and supportive measures taken if necessary.

**Duke of Argyll's tea-plant**

# · DEADLY NIGHTSHADE ·

**Atropa belladonna L.**

POISONOUS

*Fruit*
The fruit of deadly nightshade is a flat, rounded, shining black berry, 13-20mm in diameter and 7-9mm tall, carried solitary in the leaf axils on long pedicels. The berry is set above the five-pointed calyx lobes which spread out like the arms of starfish, at first green and veined, turning to the consistency of brown paper as the fruit ripens from August to October. The berries are very juicy, with an intensely sweet taste and strongly purple-staining juice. The outer part of the berry is thick and fleshy, and the two-celled fruit contains numerous blackish-brown oval seeds, 1.5-2mm long, with a beaded appearance. Salisbury calcu-

lated that the average plant carried 470 fruits, each containing some 150 seeds, with a germination rate around 60%. The seeds are mainly distributed by birds, and the plants therefore tend to grow in clumps beneath convenient perches.

*Leaf and Flower*

Deadly nightshade is a stout perennial herb, often hairy and glandular, with stout branched stems growing up to 1.5 m annually from a thick rootstock. Plants grown from seed flower from the second season. The leaves are large, entire, and oval with pointed apices and winged petioles, either alternate or in pairs of very unequal size growing together on one side of the stem. Deadly nightshade flowers from early June until late July, the drooping, bell-shaped, dirty purple flowers borne singly or in pairs on long pedicels from the leaf axils. The calyx has five pointed spreading lobes, and the flower bell has five recurved, triangular lobes. There are five stamens with yellow anthers hidden within the bell, and the single stigma just protrudes from the mouth of the flower. A form with yellowish-green flowers grows in a few sheltered habitats mainly in eastern England – it has been named var.*lutescens*. The seedlings of this variety are less hardy.

*Ecology and Distribution*

Deadly nightshade is an uncommon plant of disturbed ground on calcareous soils, particularly in woods and scrub on the Downs of Kent and Sussex, west to Wiltshire, and it is abundant in parts of the Cotswolds. Further west it grows in isolated colonies in Somerset and Devon, by the River Wye in Gwent, and on the limestone of North Wales. In the east it is found in Cambridgeshire, Suffolk, and Horsey on the Norfolk coast, where it appeared in arable land following flooding with sea water. It also grows on the chalk of Yorkshire and Lincolnshire, and on the magnesian limestone of south Yorkshire, Staffordshire and Durham, and on the carboniferous limestone of north west Lancashire and south Westmorland. It is very rare in Ireland, being recorded from Meath, Donegal and Coney Island near Shannon. Records in other places as far north as Angus in Scotland are likely to be of introduced plants, since it was widely cultivated for medicinal purposes. In south west and central Europe it is found in damp places in mountainous areas. It grows in open communities with elder and nettles and often flourishes among rabbit burrows. The rabbits eat it with impunity, yet their flesh is said to be rendered toxic to any predator. The flowers are most likely to be pollinated by bees. Seed dispersal is probably effected most commonly by birds, especially pheasants.

*History and Uses*

Deadly nightshade has been cultivated for a long time for its medicinal properties. The seeds have been found among Roman remains at Silchester in Berkshire, and the old English name for it was 'Dwale'. Linnaeus named it after Atropos, the Greek Fate, one of the three daughters of Night, who with her shears cut short the thread of life. Culpeper had no use for it as an internal medicine, but used the leaves for poultices. Gerard was more forthright in his condemnation of the plant: 'Banish therefore these pernitious plants out of your gardens, and all places neere to your houses, where children or women with child do resort, which do oftentimes long and lust after things most vile and filthie: and much more after a berry of a bright shining blacke colour, and of such great beautie, as it were able to allure any such to eate therof'. In the past, deadly nightshade was widely used externally in plasters, ointment and liniment, and internally as an antispasmodic. Preparations have long been used to treat conditions of the eye, and the effect of dilating the pupil was said to make the eyes of ladies more attractive, hence the name 'belladonna'.

**Deadly nightshade**

*Toxicology*

All parts of the plant are poisonous, especially the roots, fruits and seeds. Mature plants are more intensely poisonous. The toxins are ester alkaloids of the tropane group; the roots contain mainly hyoscyamine, the leaves a mixture of hyoscyamine and atropine, and the fruits mainly atropine with other unidentified alkaloids. These alkaloids remain stable in dried material and even when boiled. The action of the poisons does not irritate the gut so there is no vomiting serving to reject the material; the onset of symptoms may be delayed for several hours.

**Symptoms** The anti-cholinergic effect of the alkaloids first cause a dry mouth, a tight feeling in the chest, and a dry, flushed skin. The pupils are widely dilated and vision is blurred. There is a desire to urinate but an inability to do so, and the body temperature may rise as high as 109°F. The heart beat is loud and arrythmic, and the symptoms progress to excitement, hallucinations, coma and death from cardiac and respiratory failure. Humans seem to be particularly susceptible to poisoning,

and with adults a dose of twenty to thirty berries will prove lethal. With children, who are attracted to the fruits, a very small dose may be fatal. One five year-old child, who chewed and then spat out a single berry, exhibited classic symptoms of poisoning, which persisted for two days.

Animals rarely eat the plants, since they have a disagreeable odour, but poisoning has been recorded in calves. Pigs, sheep and goats can eat comparatively large quantities without ill effect, but horses and cattle are more susceptible. NPIS have recorded fifty-five human cases, with no deaths. Many of the fatal cases recorded relate to previous centuries. An unusual story of deadly nightshade poisoning was related by Humphrey Gilbert Carter, one-time Director of Cambridge Botanic Gardens. An undergraduate was arrested late one night in a Cambridge street, apparently thoroughly inebriated. Fortunately he was examined at the police station by a doctor, who noted the extreme dilation of his pupils and diagnosed acute atropinization. It transpired that the student's landlady was using a decoction of *Atropa belladonna* for an eye condition, and used the same cloth strainer to strain the unfortunate student's coffee!

There is some evidence that atropine can be transferred by bees from the flowers of deadly nightshade to the honey that they make, and that this can produce symptoms in people who eat the honey.

**Treatment** Vomiting should be produced promptly, and gastric lavage performed. **DO NOT INDUCE VOMITING WITH SALT WATER OR WITH MUSTARD IN WATER**. 5-10g of activated charcoal should then be given orally, and heart function monitored. In severe cases anticonvulsants or sedatives (not morphine) may be needed, or intravenous neostigmine given, to block stimulation of the central nervous system. In cases where symptoms are severe, medical supervision should be continued for twenty-four hours.

# · *BLACK NIGHTSHADE* ·

***Solanum nigrum* L.**

POISONOUS

*Fruit*

The fruit of black nightshade is a juicy, shining, opaque-skinned, purple black berry, spherical or slightly broader than long, and 6-10mm in diameter. The fruits are carried in drooping cymes of up to nine borne on stiff lateral peduncles below or opposite the leaf axils. Each berry has a pedicel about 7mm long, and a small green calyx which has five pointed, non-clasping lobes and which is greatly exceeded by the girth of the berry. The juice of the berry is pale mauve, and there are numerous small seeds 1.7-2.4 × 1.5mm, bean-shaped and yellowish with a pitted surface. The berries are ripe from mid-August until November. A form with berries which remain green even when fully ripe has been recorded uncommonly but widely in southern counties in England. The fruits are easily distinguished from those of the unripe black type because the cuticle is transparent and the seeds are therefore visible through it. They are distinguished from the green nightshades by the small, non-clasping calyx. The juice of the berries of this form is green.

*Leaf and Flower*

Black nightshade is an annual weed with spreading or erect branched stems up to 70cm long and pointed ovate leaves with entire, sinuate or toothed margins and a winged petiole. The habit, size, leaf-form and degree of winging of the petiole are

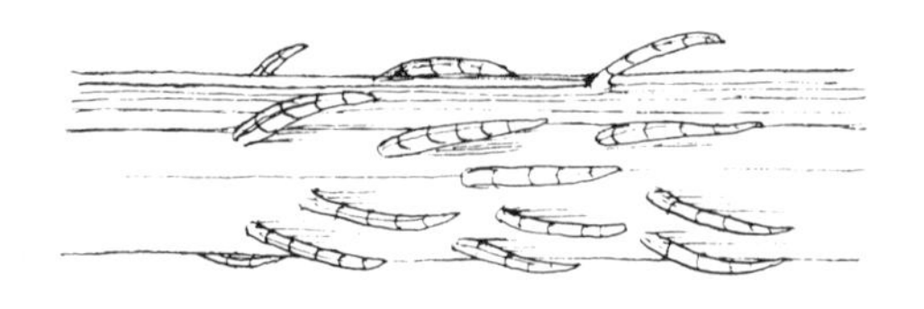

very variable, depending upon the prevalent grazing conditions and on the genetic 'strain' of the plants concerned. Grazing and mowing will reduce plant stature markedly. Most plants have leaves which are stiff and crisp, with a few hairs scattered on the surface, especially along the ribs of the leaves and the angles of the stems. In the commonly encountered form the stem hairs are mainly appressed and eglandular.

An alien race, *S.nigrum* L.ssp.*schultesii* (Opiz.) Wessely from the drier part of central-southern and east Europe is established in a number of sites in south and east England, although it is still comparatively rare. It has been found in Southampton docks, near Weybridge in Surrey, in four sites in north west Kent, in several sites in Essex near Rainham and Hornchurch, and at Holme in West Norfolk. It is distinguished by having hairy stems, with some of the hairs ascending or spreading, most hairs having a small apical gland which is ovoid and glistens in bright light.

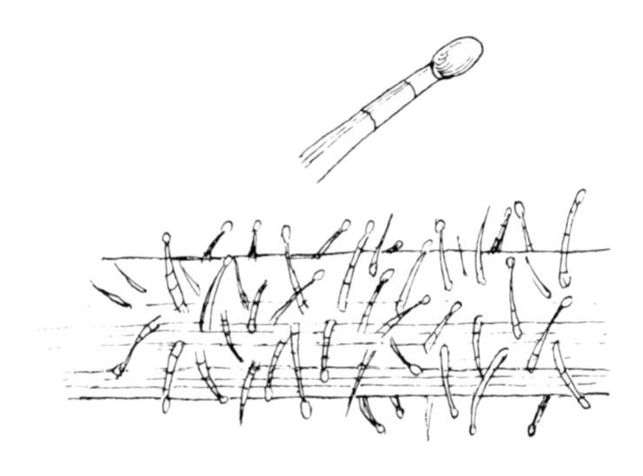

Black nightshade flowers from early July until October, with up to twelve flowers, 8 mm in diameter, in short dense leafless cymes on a peduncle below or opposite to the leaf attachment. The calyx has five small blunt lobes, and the corolla is white, with five pointed petals like a star, which may fold back at full maturity, closing over the anthers later. At the centre is a prominent cluster of five yellow anthers. Some flowers may be tinged with purple.

*Ecology and Distribution*

Black nightshade is a common plant of cultivated and disturbed land in England, especially in the south and east. In the south west it is common in Somerset, except in the hills, and it is widespread in Wales, especially in the south. It grows fairly commonly as far north as Cumbria and north Yorkshire, but is absent from Scotland except as a rare casual. It is similarly rare in Ireland, recorded in north east Galway, and as an abundant introduction in tilled fields around Kilgobbin and Foxrock in Co Dublin. It grows well between the rows of vegetable crops and around farmyards and slurry pits rich in nitrogenous waste. In Europe it is widely distributed, but doubtfully native, in most of the northern countries.

Seed production is prolific, the average plant yielding about 1500 seeds, and the seeds are extremely viable. In experiments reported in the Journal of Agricultural Research in 1946, 83% of seed proved viable after thirty-nine years. This helps to explain the sudden appearance of masses of plants in newly cultivated ground.

*History and Uses*

Seeds of black nightshade have been found in Palaeolithic deposits in Britain, and the predelection of the species for sites rich in nitrogen is reflected in its Saxon name 'Mixplenton' – the plant of the mixen or dung heap. It has been used externally in liniments and poultices. Culpeper noted: 'It is used to cool hot inflammations, either inwardly or outwardly, being in no way dangerous, as most the Nightshades are'. The ripe fruits are sometimes eaten in Africa and Australia, while vegetative shoots are also eaten in parts of Europe, Africa and Asia.

*Toxicology*

Most authorities state that while the ripe berries are more or less innocuous the unripe fruits and all green parts of the plant are toxic, containing the glycoalkaloid solanine, with the aglycone solanidine, the alkaloids tropeine and tropane, nitrates and nitrites. The action of these substances is both irritant to the stomach and depressant to the central nervous system. Assay of plants and berries, both in Britain and in Australia (New South Wales), where the species is introduced and regularly fruits, usually fail to demonstrate any poisonous alkaloids. However, in view of the various races and forms of black nightshade which grow in Britain, it would appear that under certain conditions of soil and climate the plant can be toxic.

**Symptoms** Poisoning in humans causes nausea and

vomiting, with profuse diarrhoea and colic. There is headache, with mental and respiratory depression, a cold clammy skin or fever, and delirium with a slow or fast pulse. NPIS have recorded poisoning commonly, the majority of cases involving accidental ingestion by children, but severe symptoms are seldom seen, and deaths are rare. It has been reported as causing poisoning in pigs, producing perirenal oedema, and in goats and cattle, which had laboured respiration, constipation followed by diarrhoea, and dry, cold extremities. In the dry summer of 1976, when forage for cattle was in very short supply, a herd of friesian cows in East Sussex were fed lucerne heavily contaminated with black nightshade. Each cow consumed an estimated 5kg of leaves and berries daily for six weeks without ill effect.

**Treatment** Induce vomiting and use gastric lavage only if large quantities have been eaten. Fluid replacement therapy will be needed where diarrhoea is severe, and the patient should be kept warm, or cooled in the occasional cases where fever is a symptom.

**Black nightshade**

# · GREEN NIGHTSHADE ·

***Solanum nitidibaccatum*** **Bitter and *S.sarrachoides* Sendtner**

POISONOUS

Several alien members of the *Solanaceae* have established themselves locally in southern Britain. These two green nightshades are well established and appear to be spreading. Much of the collected material is attributable to *S.nitidibaccatum*, which would appear to be the more common species in Britain.

## *Solanum nitidibaccatum*

*Fruit*

The fruit of this green nightshade is a globular berry 6-9mm in diameter, dark green or brown to purplish brown and translucent when ripe, so that the seeds and internal divisions are visible through the cuticle. They are carried in drooping cymes of four to eight on peduncles usually opposite the leaves. The stiffly hairy calyx, with five pointed lobes, partly clasps the basal one-third to one-half of the fruit. The berries are shiny, with minute rough projections on the surface, and very juicy. They are ripe from September onwards and contain fifteen to twenty-four brown seeds, 1.8-2.4 × 1.3-1.9mm, having a pitted surface, together with one or two small sclerotic granules, less than 0.8mm in diameter.

*Leaf and Flower*

*S.nitidibaccatum* is a low, bushy, pale-green annual with stiff, ribbed stems and crisp, pointed oval, toothed leaves, all densely hairy with stiff, multiple-celled, glandular hairs. The flowers are 8-9mm in diameter, with five pointed, stiffly hairy calyx lobes, five pointed white or purplish corolla lobes with a distinct purple and yellow basal star, and a cone of yellow stamens in the centre. Flowers are present from July to late September.

*Ecology and Distribution*

*S.nitidibaccatum* is an alien from Chile and Argentina which is most often found in market gardens – in crops of carrots or beet – on light loam or sandy soils. It has been widely recorded from south east and east England, but the true distribution picture is unclear because of confusion with the following species. It is also found as a casual in various European countries.

***Solanum nitidibaccatum***

## *Solanum sarrachoides*

*Fruit*

The berries are similar to those of *S.nitidibaccatum*: translucent light green, 6-9mm in diameter – but

the calyx lobes are strongly accrescent, cupping round the ripe berry so that it is almost completely hidden. The berries contain fifty-nine to sixty-nine pale yellow, blunt ended seeds, 1.3-1.5 × 1.0-1.3 mm, with a pitted surface. They also contain four to six large sclerotic granules, 1-1.3 mm across. The fruits are ripe from September onwards.

*Leaf and Flower*

*S.sarrachoides* is a low, bushy green annual, much softer and more floppy than *S.nitidibaccatum*. The leaves are larger, pointed oval, scarcely indented, and softer in texture, with scattered, long glandular hairs, mainly on the petiole and ribs of the leaves. The flowers are borne in drooping cymes of four or five, and are white, with an inconspicuous basal star. The calyx bears long, soft multicellular hairs and the calyx lobes are partially united in a deep cup; it is impossible to set them out flat without splitting them. This species flowers from July to September.

*Ecology and Distribution*

*S.sarrachoides*, like *S.nitidibaccatum*, is an alien weed on cultivated land, and has a similar distribution, mainly in south east and east England. It is particularly well distributed in Bedfordshire and Cambridgeshire. It appears to be increasing in west Suffolk, and has been recorded west to Gloucestershire, Shropshire and North Wales. The pattern of distribution is unclear, and it would seem to be rarer in the British Isles and Europe than *S.nitidibaccatum* (some previous records are incorrect).

*History and Uses*

Culpeper described a green-berried nightshade (*Solanum nigrum baccis viridis*) with indented leaves, greenish-yellow berries and a low-branching habit. It is likely that he was referring to the green-berried form of *S.nigrum*. Although there is little evidence it would be wise to assume that these two species of green nightshade are toxic, and to avoid them.

*Toxicology*

NPIS have recorded no cases of poisoning in humans. Suspected poisoning should be treated in the same way as for black nightshade.

***Solanum sarrachoides***

# · BITTERSWEET ·

**Solanum dulcamara L.**

POISONOUS

*Fruit*

The fruit of bittersweet is a brilliant shiny red berry, ovoid, 9-10mm long and 8mm in diameter, carried opposite the leaves in drooping, much-branched cymes of ten to twenty on stiff pedicels 7-8mm long. The berries are thin-skinned and fleshy, packed with numerous yellow seeds, 2.5mm long, lenticular with a finely pitted surface. They are very juicy, with an intensely bitter taste which leaves a sweet aftertaste in the mouth. The berries are ripe from early September and can remain on the plants until January of the following year.

**Bittersweet**

*Leaf and Flower*

Bittersweet is a much-branched, climbing, weak-stemmed perennial which may grow to 3m in hedges. The stalked upper leaves are often lanceolate while the lower leaves are spear-shaped with two separate lateral lobes at the base. The flowers, which appear from mid-June to September, are borne on peduncles opposite the leaf-stems. The calyx has five small, blunt lobes,and the corolla is a showy purple, with five pointed lobes, each with a yellow spot at the base. The lobes may be reflexed. The five stamens, which have yellow anthers, form a projecting cone in the centre of the flower. Plants with white flowers are recorded rarely. On coastal shingles in southern England a prostrate form of

bittersweet, *S.dulcamara* var.*marinum* Babington is found. This variety grows as a small prostrate bush with persistent woody stems of moderate length and berries of a more globular shape. In the windy, salt-laden atmosphere, fruiting is seldom as heavy as on inland plants, and the berries soon shrivel and fall. Seedlings of this variety, transplanted at the cotyledon stage and grown inland in good soil retained their prostrate woody growth habit for at least two years of cultivation, and fruited in the first season.

*Ecology and Distribution*

Bittersweet is a common plant of hedges, river banks and sandy places throughout Britain north to Sutherland, and on coastal shingles in the south of England. It is most frequent in the southern half of Britain, becoming increasingly rare further north in Scotland, and being absent from the Hebrides, Orkney and Shetland. In Wales it is absent from the mountainous areas, and in Ireland it is sparsely scattered. Like all solanaceous plants it flourishes best in a warm environment. It is found in similar sites throughout most of Europe, but is absent from the extreme north.

*History and Uses*

Bittersweet has been known and used medicinally for a very long time. The scientific name is an inversion of the mediaeval Latin *amaradulcis* – bitter sweet. Culpeper wrote: 'Country people used to take the berries of it, and having bruised them, they applied them to felons, and thereby soon rid their fingers of such troublesome guests' – a felon being a small abscess or whitlow, hence the old name of Felon-wort. Gerard recommended: 'The juice is good for those who have fallen from high places, and have been thereby bruised, or dry-beaten'. As late as 1934 Dulcamara, composed of the dried stems and branches, was listed in the British Pharmacoepia, and, used as an infusion, was a popular remedy for chronic rheumatism and skin eruptions. Preparations containing bittersweet are still used in homoeopathic medicine.

*Toxicology*

All parts of the plant are poisonous, the unripe fruits being particularly dangerous. The main principles are the alkaloids solanine, solaneine and solaceine, with the saponins dulcamaric and dulcamaretic acid which are concentrated in the berries. These substances are irritant to the stomach and intestine, affect the central nervous system when taken in very large quantities, and have a narcotic action.

**Symptoms** Children are attracted to the fruits, as they are bright-red and shiny. Symptoms of poisoning may be delayed for some hours. There is drowsiness and a staggering gait, then a harsh, itchy feeling in the mouth and pharynx and finally nausea, vomiting, colic and copious dark-coloured diarrhoea in which undigested red berries may appear. The mouth is not dry. The pupils are normal and react normally to light. The patient feels dizzy, weak and trembling, and in severe cases respiration is depressed, with a slowing of heart rate and drop in body temperature. NPIS list twelve cases, which exhibited mild symptoms only and recovered fully within twenty-four hours. Livestock rarely eat bittersweet, but some affected animals become addicted to it. Pigs recovering from poisoning showed evidence of dermatitis.

**Treatment** Vomiting should be induced, and 5-10 g activated charcoal left in the stomach after gastric lavage. Support for the cardiovascular system and respiration may be necessary, and stimulants such as warm, strong, sweetened tea should be given to drink.

# · *WILD MADDER* ·

***Rubia peregrina*** **L.**

*Fruit*

Wild madder is unusual among *Rubiaceae* (Bedstraw Family) in having a fruit which is a succulent berry. The berries of wild madder are usually globular, 4.5-6 mm in diameter, with a shiny-black, leathery skin which is finely pitted. The berries are borne in axillary cymes of ten or more, on reddish sharply angled pedicels from which they readily detach when ripe. The apex of the berry is often at 90° to the point of attachment of the pedicel, and most berries contain a single, round, black seed 3.5 mm in diameter, with a broad white base. Occasionally a berry will be found containing two seeds, the berry being broader than long, and in this case the pedicel attachment is always opposite to the floral remnants. The berries contain a dark purple-staining juice and are ripe from mid-October, often lasting on the plants over winter, as late as April of the following year.

*Leaf and Flower*

Wild madder is a stout, straggling, branched perennial herb, often evergreen, with the lower part of the stem woody and persistent, growing up through other bushes to 2 m. The stem is sharply four-angled, with hooked, downward-pointing, bristles on the angles. The leaves are in whorls of four to six, broad ovate with a point at the tip, shiny, thick and rigid, with strong downwardly directed prickles on the margins and on the underside of the midrib. The flowers appear from mid-June to August, with typical bedstraw-like blooms 4.5-6.5 mm long, in spreading, long-stalked cymes. The calyx forms a ring and the corolla is yellow-buff, with five pointed lobes. The five stamens alternate with the petals, and the stigma is bifid.

*Ecology and Distribution*

Wild madder is a mainly sub-maritime plant, growing in scrubland on cliffs and in hedges near the sea, where it can scramble extensively over bushes and rocks. It reaches its eastern limit on the cliffs of Dover and Deal, where it is rare. In Sussex it is a rare plant growing in several locations on the chalk near Arundel. It is plentiful on the cliffs of the Isle of Wight and Dorset, and all along the south west coast to Cornwall. Near Lostwithiel it grows on sea cliffs with *Asplenium marinum* and *Orobanche hederea*. It is found in many sites around the coast of Wales, particularly in the south from Cardiff westwards to St Anne's Head in Pembrokeshire. It is absent from the whole of the east coast of England and from Scotland. In Ireland it occurs all along the south coast from Waterford to Kerry, and

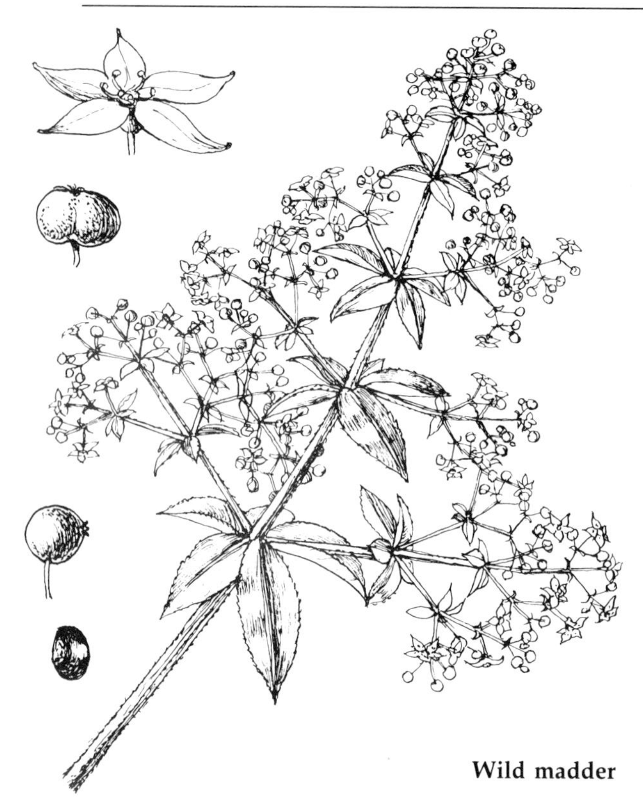

**Wild madder**

on the west coast particularly around Galway Bay and on the coast of the Burren. It is rarely found inland, but in Somerset it is not uncommon both on lias and limestone away from the coast. In Europe it is found in the south and west, northwards to 53°30′, growing in hedges, thickets and on rocky ground. It is also part of the flora of maquis and garrigue around Mediterranean shores as far east as Israel.

# · *DWARF ELDER* ·

***Sambucus ebulus*** **L.**

*Fruit*
The fruit of dwarf elder is a small, globose drupe, 4-5mm in diameter, in clusters of as many as a hundred on reddish pedicels of varying length, so that the erect fruiting head is almost flat-topped. When ripe the fruits are purplish-black, not very shiny, and crowned with the rough stump of the floral remnants. They closely resemble the berries of elder, but the umbel is erect, not drooping, and consists of three main branches, instead of the five

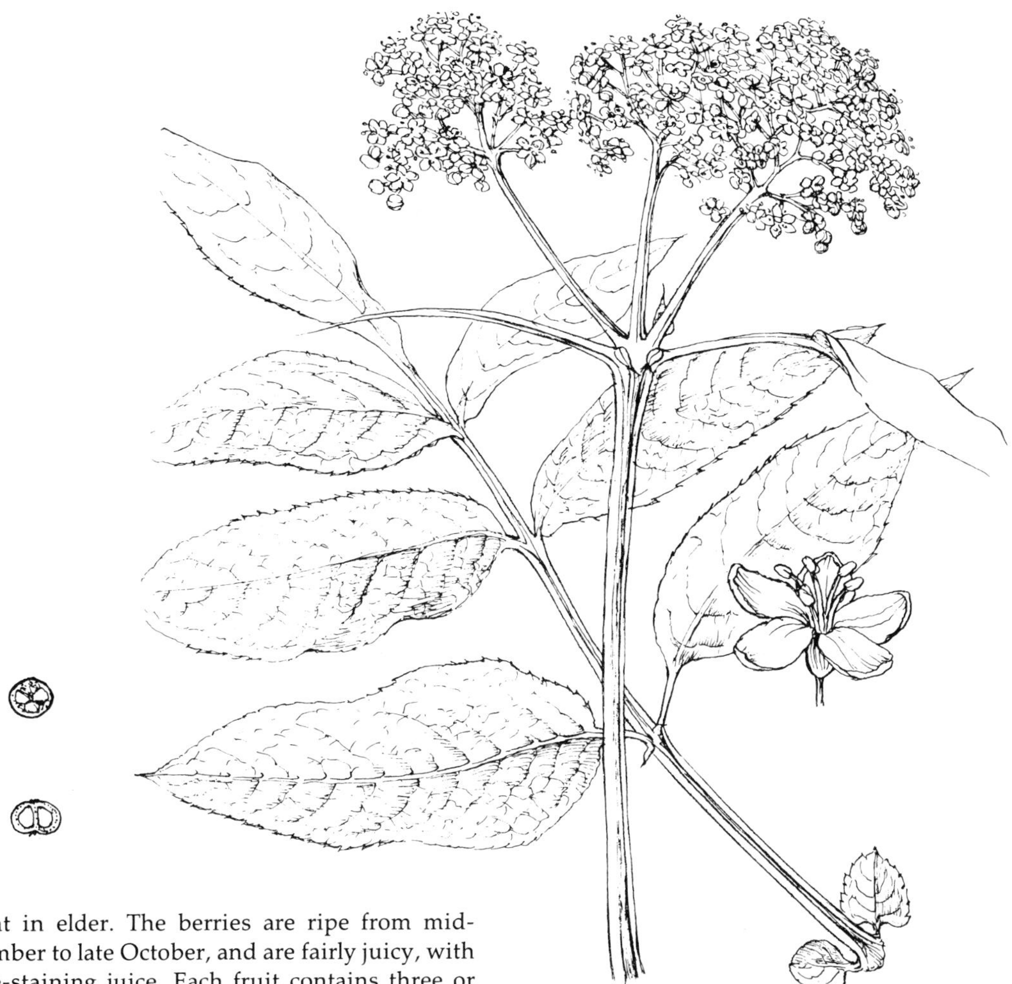

present in elder. The berries are ripe from mid-September to late October, and are fairly juicy, with purple-staining juice. Each fruit contains three or four pyrenes, 2.5-3 × 1 mm, triquetrous and yellow, with a netted, wrinkled surface on the outer curved side of the seed. The berries are sweet, and taste much like elder berries.

*Leaf and Flower*

Dwarf elder is a stout, glabrous perennial, with grooved, simple herbaceous stems growing to above 2 m each year from a woody, creeping base. They die back underground in winter time, forming persistent patches which spread slowly. Aerial shoots do not appear until early summer but then growth is rapid. The leaves are pinnate with four or five pairs of lanceolate, finely-toothed leaflets and broad ovate toothed stipules at the base. The growing plant much resembles a stout umbellifer and the foliage has a strong and disagreeable smell. Flowers are produced in dense flat-topped corymbs from mid-July to late August, each flower with a corolla of five creamy, pink-tinged petals, and stamens with strongly contrasting purple anthers.

*Ecology and Distribution*

Dwarf elder is a rare plant with a scattered distribution throughout Great Britain north to Sutherland and Caithness and west to the Burren of western Ireland where in some places it grows lush and tall. It is a plant of wasteland, field edges and road verges, mainly on limestone soils, principally in the southern half of England. It is never common – having probably been introduced in many of its sites – but persisting for many years where it is established. It occurs in seven stations in Kent, ten in Sussex and twelve in Somerset; in Wales it is well represented in all areas as far west as the Lleyn Peninsula. It is decreasing in Cumbria and Suffolk, where it still occurs in nine sites. It has disappeared recently from Angus. In Europe it grows in most countries, from Holland and the north Ukraine southwards. As it has been cultivated for medicinal purposes it is widely naturalized.

*History and Uses*

Legend relates the dwarf elder (Danewort), grew from the blood of Englishmen killed in battle by the Danes. The name may have derived from 'dain', to stink – from the unpleasant smell of the foliage. The Old English name for it was Walwort – Foreigner's Plant. The first written reference was in 1491 by John Ross, an antiquary of Warwick, and the first use of the name Danewort was by Turner in 1538. Culpeper wrote that dwarf elder: 'is more powerful than Common Elder in opening and purging Choler, Phlegm and Water . . . The berries are likewise sudorific . . . as an extract called Rob of Elder. I would recommend to the patient to take half a dram of fine levigated crab's eyes, mixed up with half a spoonful of water, and immediately after six drams or an ounce of the said Rob'. Gent noted: 'Dwarf Elder is of a Temperature hot and dry in the third degree, it doth waste and consume by Purging of Choler, and Phlegm, and Water, and is more powerful than the common Elder, and hath all the Properties of it.' Deakin described the root and whole plant as possessing violent purgative and emetic properties, and said that it was prized in Saxon times as an efficacious treatment for dropsy.

William Woodville, in his *Medical Botany* of 1790, also noted these qualities in preparations of the root. He wrote: 'The berries, in their recent state, according to Scopoli, prove a gentle cathartic, though Haller says that he never experienced this effect from their use. The seeds are said to be diuretic, and to have been given with advantage in dropsical complaints; they also afford an oil, which Haller applied with success in painful affections of the joints . . . The odour of the green leaves drives away mice from granaries; and the Silesians strew these leaves where their pigs lie, under a persuasion that they prevent some diseases to which these animals are liable.'

*Toxicology*

The berries and roots contain a saponin, resin, bitters and traces of a cyanogenic glycoside, but the toxicity is very low and there have been no recent authenticated reports of poisoning in humans or animals. Cows have been known to eat the foliage and berries with apparent impunity.

**Dwarf elder**

# · ELDER ·

**_Sambucus nigra_ L.**

POISONOUS

*Fruit*

The fruit of elder is a berry-like drupe, globular, 6-8mm in diameter, and shiny black in colour. The berries are borne in dense, level-topped corymbs on reddish pedicels, the corymb having five main rays, measuring 20cm or more in diameter, and often containing more than a hundred berries. When the elderberries are ripe in late August and September the clusters of fruit droop down. The fruit is sweet and juicy, with dark purple-staining juice and a tang which is slightly irritant to the throat if the berries are eaten raw. Each berry contains three to five flattened, hard pyrenes 2.5 × 1mm. Elder bushes with berries which remain green when ripe, *S.nigra* var.*viridis* Weston, occur rarely and may be introduced.

*Leaf and Flower*

Elder is a deciduous shrub or small tree growing to 7m, with a soft, fissured bark and stems full of pith. It suckers readily from the base. The leaves are dark green, pinnate, with five broad ovate leaflets which are pointed and finely toothed. Stipules may be present but, if so, they are very small and simple. The young green twigs bear numerous scales. Flowers appear from late May to early July, the creamy florets, each 5mm in diameter, borne in dense, flat-topped corymbs. Each flower has five petals, five creamy anthers and three to five stigmas. The scent is very strong and cloying, attracting numerous small flies which act as pollinators. A form of elder with rather feathery, pinnate leaves, var. *lacinata*, occurs occasionally in Kent, Sussex, Suffolk and as far north as Angus, and may be planted in shrubberies.

*Ecology and Distribution*

Elder is a very common plant of woodland, hedgerows and scrub throughout the British Isles. It is an important element of chalk scrub, and is found

**Left: Elder**
**Right: Green-berried elder**

particularly on disturbed ground near rabbit warrens. The rabbits leave it alone as they dislike its taste. It is less common in the north west of Scotland and the Highlands, and is an introduced species in Orkney and Shetland. It is common throughout Europe except the extreme north.

Two other species of elder have been introduced and may establish themselves locally. Red-berried elder (*S.racemosa*) is distinguished by its greenish-yellow flowers and red berries. It comes from Europe and western Asia, and has been recorded in Kent, Norfolk, Suffolk, Breconshire, Radnorshire, Cardiganshire, Montgomeryshire, Westmorland and parts of east Scotland. *S.canadensis* has become naturalized on railway banks and in gardens. It is a bushy plant with large, bright green pinnate leaves with seven leaflets, large hummocky clusters of creamy flowers, and purplish fruits.

*History and Uses*

Elder has been used and cultivated for many centuries. Legend has it that the Cross was made of elder wood and that Judas Iscariot hanged himself from an elder – though he would have found it difficult, as the branches break and tear off very easily (for this reason people in the past were loth to cut or burn it). It was also planted as a protection against lightning and against witches. In Pliny's time elderberry juice was used as a hair dye. Culpeper noted: '. . . the berries, either green or dry, expel the same humour, (vis. phlegm) and are often given with good success to help the dropsy. The juice of the berries boiled with honey, and dropped into the ears, helps the pains of them; the decoction of the berries in wine being drunk, provokes urine'. Gent likewise observed: 'Elder is of a drying quality, glewing and moderately digesting; it purgeth choler and phlegm, both the inward Rind and the Berries'. Elder flowers are widely used for making wine; as a delicious additive to gooseberry jam; and in elder flower fritters. The berries also make excellent wine, and can be used in a wide variety of puddings, jellies and preserves. Thirty-two species of birds have been recorded eating the berries between August and December, especially blackbirds, song thrushes, mistle thrushes, ring ouzels, starlings and green finches. A male greenfinch which has over-indulged in elderberries and ended up with a mauve head can create problems for bird-watchers!

*Toxicology*
Leaves, stems, bark, roots and berries are all mildly toxic, containing terpene volatile oils, and glucosides rutin, sambunigrin and quercitrin, irridoids, anthocyanins and tannins. The leaves, bark and unripe fruits contain cyanogenic glycosides. Poisoning can cause mild gastro-intestinal irritation, especially if unripe or under-cooked berries are eaten in large quantities.
**Symptoms** There will be nausea, and possibly vomiting with mild purgation. There may be dizziness and accelerated heart rate, and even convulsions, but such symptoms are rare. NPIS list fewer than ten cases. Elder appears unattractive to most animals. According to a report from Romania, pigs poisoned with the leaves salivated profusely and exhibited diarrhoea and vomiting, with an increased pulse and repiratory rate, trembling and ataxia. Fourteen out of a group of fifty affected pigs subsequently died.
*Treatment* Symptomatic care may be needed. If a child has eaten large numbers of berries, induce vomiting with syrup of ipecacuanha. Fluids may need to be replaced orally.

# · GUELDER-ROSE ·

**_Viburnum opulus_ L.**

*Fruit*
The fruit of Guelder-rose is a globular drupe, each containing a single pyrene. It is carried in a broad corymb of twenty to thirty, the individual berries on very short pedicels 1-2mm long, and the whole cluster drooping when they are ripe in September and October. Ripe fruits will stay on the bushes well into November, even after the leaves have dropped. They are highly attractive: a clear, shiny translucent scarlet, each 7-10mm in diameter, and very juicy. A yellow-fruited form was found near Gomshall, Surrey in 1946. The single, buff-coloured pyrene is relatively large, 6-7mm long, 6-7mm wide, pointed ovate and flat, with the radicle forming a pronounced rib from base to point of the seed. The flavour of the ripe fruit is rather unpleasant.

*Leaf and Flower*
Guelder-rose is a large deciduous shrub up to 4m high, with ovate leaves which are divided into three to five large triangular-toothed lobes. In au-

**Guelder-rose**

tumn the leaves turn a beautiful flaming red. Flowers are borne in broad, erect corymbs, consisting of an outer ring of irregular enlarged flowers, 15-20mm in diameter, which are sterile. The central flowers are regular, 6mm in diameter and fertile. The corolla is whitish, five-lobed, and slightly fragrant. The flowering period is from early June to mid-July.

*Ecology and Distribution*

Guelder-rose grows in woods, scrub and hedgerows, preferring moist clay soils, and occurring fairly commonly throughout England, Wales and Ireland. Although it may occur in chalk scrub it tends to avoid chalky and light soils. In East Anglia it is found in alder carr, fens and damp scrub. Although it may be found in Scotland as far north as Sutherland and Caithness it is nowhere common and is absent from the Outer Hebrides, Orkney and Shetland. In Europe it is widespread except for parts of the north and most of the Mediterranean.

*History and Uses*

The name Guelder-rose originated from the town of Guelders which lay on the border between Prussia and the Netherlands. It was probably first used to describe the cultivated form of *V.opulus*, the snowball-tree, which has a round flower-head. Gerard mentioned it in his Herbal (1597): 'The Rose Elder is called in Latine *Sambucus Rosea* and *Sambucus aquatica* . . . in English Gelders Rose and Rose Elder'. When raw the berries are not really palatable but they are safe when cooked. An infusion of the bark was used to treat cramp, hence the name 'Cramp Bark' which was sometimes used.

*Toxicology*

The berries of Guelder-rose are of very low toxicity but may cause vomiting if eaten raw. They contain tannin, resins, valerianic acid and a glycoside which has been called viburnine.

# WAYFARING-TREE

**_Viburnum lantana_ L.**

*Fruit*

The fruit of wayfaring-tree is an oval drupe 7-10mm long and 5-6mm wide, with a shallow groove down each side, borne in a dense, slightly domed corymb. When the fruits are under-ripe they are brilliant scarlet, each with the dark, shrivelled floral remnants at the top. They turn jet black when ripe from early August until October. The flat heads with a mixture of black and red berries make identification of the fruiting bushes easy. The skin is tough and shiny, the flesh rather slimy in texture, brownish black in the ripe fruit, with a sweet, faintly plum-like flavour. Most fruits contain a single flat ovate seed, 7mm long and 5mm wide, brown and rough textured, with three grooves on one face, and the radicle forming a central rib with a groove either side of it, on the other face.

*Leaf and Flower*

Wayfaring-tree is a deciduous shrub growing to 5m, with downy, brown, rounded twigs and blunt oval, opposite leaves, grey-green in colour, with a wrinkled surface and finely toothed margins. The leaves are also downy, especially on the underside along the veins and midrib, the hairs being very small and star-shaped. Flowers appear from late April until early June, and occasionally bushes will be found in bloom as late as October. The flowers are borne in dense corymbs 6-10cm across, with no showy sterile flowers on the periphery. Each flower, 5-8mm in diameter, has five very small calyx teeth and a campanulate white, five-lobed corolla, from which project five long yellow stamens. The flowers have a rather sickly fragrance. In autumn the leaves turn beautiful shades of mauve and red.

*Ecology and Distribution*

Wayfaring-tree is most common in southern England, south of a line from the Wash to the Severn, where it is a common plant of woods, scrub and hedgerows on basic soils, and an important ele-

**Wayfaring-tree**

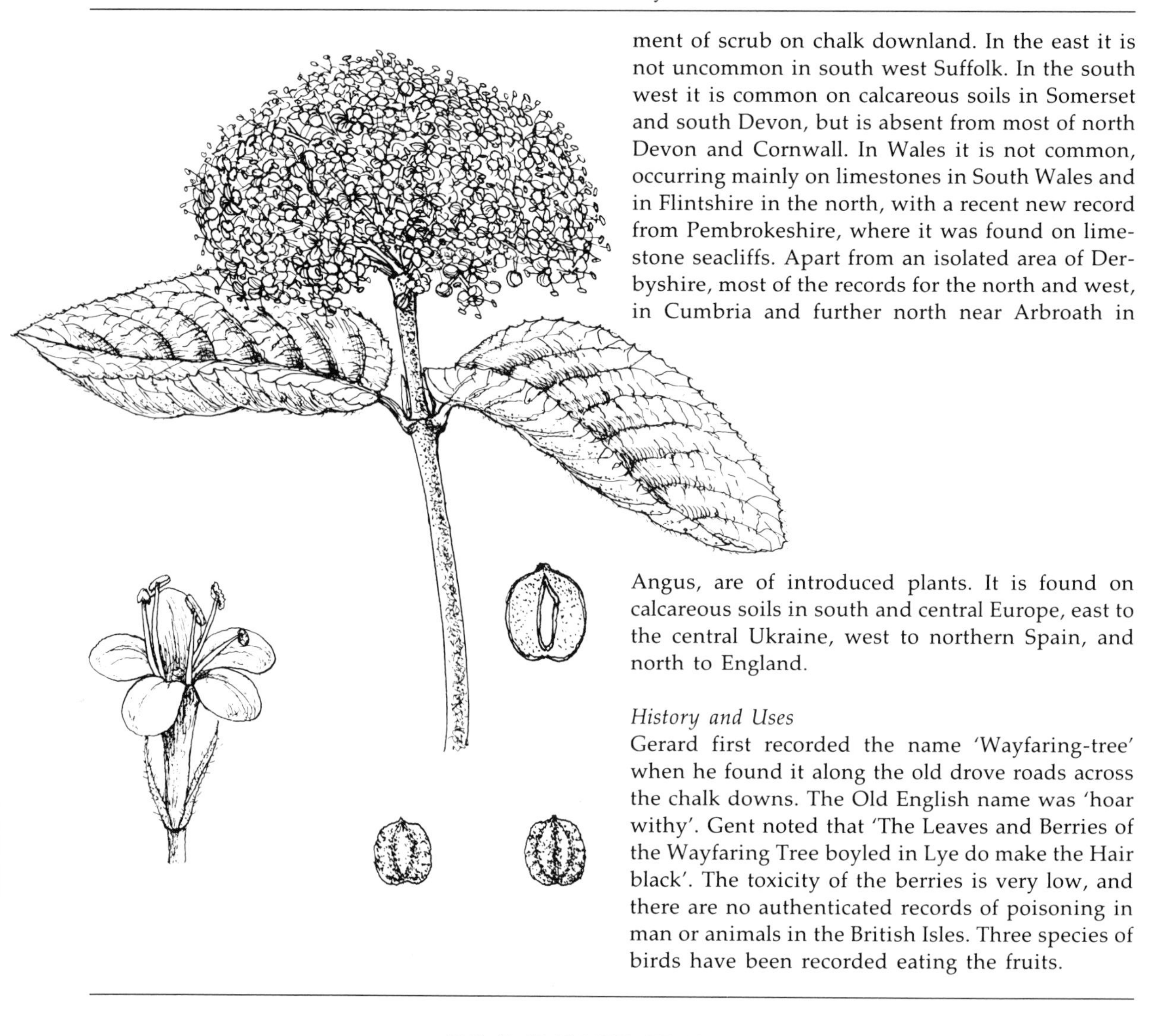

ment of scrub on chalk downland. In the east it is not uncommon in south west Suffolk. In the south west it is common on calcareous soils in Somerset and south Devon, but is absent from most of north Devon and Cornwall. In Wales it is not common, occurring mainly on limestones in South Wales and in Flintshire in the north, with a recent new record from Pembrokeshire, where it was found on limestone seacliffs. Apart from an isolated area of Derbyshire, most of the records for the north and west, in Cumbria and further north near Arbroath in Angus, are of introduced plants. It is found on calcareous soils in south and central Europe, east to the central Ukraine, west to northern Spain, and north to England.

*History and Uses*

Gerard first recorded the name 'Wayfaring-tree' when he found it along the old drove roads across the chalk downs. The Old English name was 'hoar withy'. Gent noted that 'The Leaves and Berries of the Wayfaring Tree boyled in Lye do make the Hair black'. The toxicity of the berries is very low, and there are no authenticated records of poisoning in man or animals in the British Isles. Three species of birds have been recorded eating the fruits.

# · SNOWBERRY ·

***Symphoricarpos albus* (L.)S.F. Blake**

POISONOUS

*Fruit*

The fruit of snowberry is a fat, white drupe, irregularly globose, 10-15mm in diameter, borne virtually sessile in clusters on the ends of the drooping branches. Each cluster may have as many as twelve well-developed berries and a number of shrivelled abortive ones. Each fruit has a thin, dull coat, the apex of the berry being crowned with small, black, shrivelled floral remnants. The flesh is like soft meringue, with a faintly sweet taste, and it contains two well-developed loculi and two which fail to mature. Each fertile loculus contains a seed

**Snowberry**

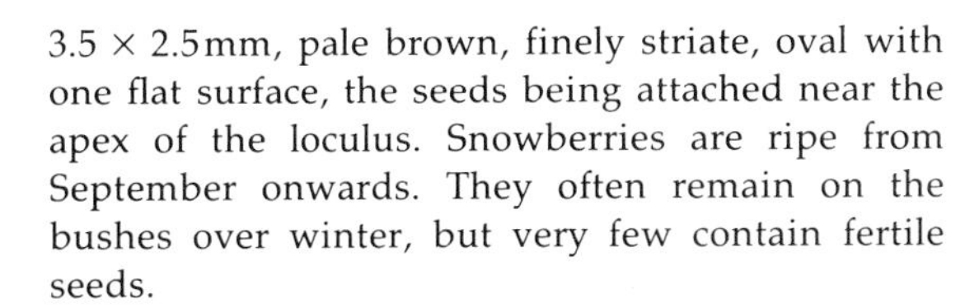

3.5 × 2.5mm, pale brown, finely striate, oval with one flat surface, the seeds being attached near the apex of the loculus. Snowberries are ripe from September onwards. They often remain on the bushes over winter, but very few contain fertile seeds.

*Leaf and Flower*
Snowberry is a little-branched, deciduous shrub 1-3m tall with slender glabrous branches; the British form appears to be *S.albus* var.*laevigatus* (Fernald) S.F. Blake. The leaves are small, ovate and opposite on slender petioles, and may have slightly wavy edges. The small flowers are produced from mid-June to early September in short, crowded axillary racemes of a dozen or more. They are 4-6mm in diameter, the calyx having four or five minute teeth. The campanulate pink corolla has five lobes and five stamens.

*Ecology and Distribution*
Snowberry was introduced into the British Isles in 1817 from North America, and not surprisingly this attractive shrub was planted extensively in shrubberies as hedging. It has become naturalized and has spread extensively, especially by streams and damp ditches; in one such typical site between

Coniston and Kettlewell in Wharfedale it is magnificent and lush, growing as a long hedge. It has spread and increased greatly in recent years in the south of England in Kent and Sussex. It is widespread everywhere except in the north of Scotland, though it has been recorded as far north as Kildonan and Tongue in Sutherland. It is absent from the Hebrides, Orkney and Shetland. Three species of bird have been recorded eating the berries.

*Toxicology*
The fruit is poisonous, containing saponins and other unidentified irritants, alkaloids which may be viburnine or chelidonine, tannins, terpenes, coumarins and triglycerides. These are irritant to the stomach and also to the skin.
**Symptoms** Poisoning will cause vomiting, mild dizziness and possibly drowsiness. If large quantities are eaten there may be profuse vomiting and protracted diarrhoea leading to dehydration. Lampe and Fagerström in America report that the berries cause vomiting and drastic purging. In one case involving four children consumption of the berries led to delirium, followed by semi-coma. These effects may persist for twenty-four hours. The juice of the berries can cause dermatitis. NPIS have recorded five cases in Britain, all with minor symptoms, which made an uneventful recovery. No animal poisoning has been recorded.
**Treatment** Vomiting should be induced if more than three or four berries have been eaten by a child or more than ten by an adult. Demulcents should be given orally to soothe the gut, and oral fluids should be provided to prevent dehydration. Antihistamine creams should be used in cases where dermatitis is present.

# · *FLY HONEYSUCKLE* ·

***Lonicera xylosteum* L.**

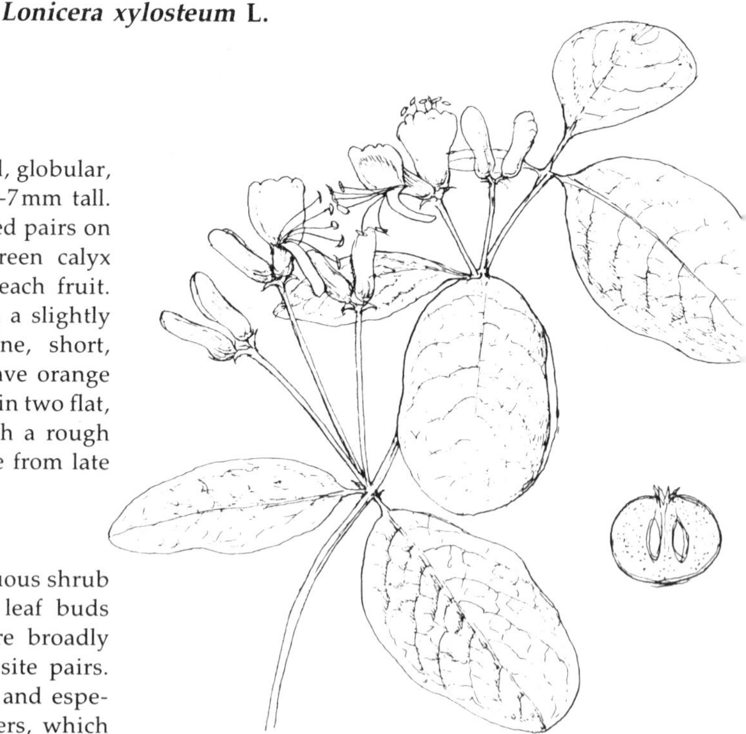

*Fruit*
The fruit of fly honeysuckle is a flattened, globular, juicy red berry 8mm in diameter and 6-7mm tall. The berries are borne in closely appressed pairs on pedicels 15mm long, the two small green calyx remnants hidden beneath the base of each fruit. The berries are translucent scarlet, with a slightly frosted appearance created by the fine, short, appressed hairs on the surface. They have orange flesh and sticky juice. Most of them contain two flat, oval irregular seeds, 3-3.5mm long, with a rough surface and yellow colour. They are ripe from late July to mid-September.

*Leaf and Flower*
Fly honeysuckle is an erect, bushy deciduous shrub 1-3m tall with delicate branches. The leaf buds open in early April, and the leaves are broadly ovate, entire and short-stalked in opposite pairs. They are grey-green, softly hairy above and especially beneath. The creamy yellow flowers, which are produced in axillary pairs on a shared stalk, appear from early to late May. The corolla is tubu-

lar, 8-12mm long, and two-lipped. The lower lip is simple and hairy outside; the upper lip is four-lobed. The flowers are unscented, but are pollinated by bumble bees.

*Ecology and Distribution*
Fly honeysuckle is a very rare native, known for a long time from the chalk near Amberley in Sussex. It has been recorded from Wilmington in East Sussex since 1901, and may be native there also. Elsewhere it is dubiously native, but has been recorded widely in England, Wales, Scotland and Ireland, although it is nowhere common, the most recent new record coming from Cumberland. In

**Fly honeysuckle**

Sussex it grows under ash and whitebeam on a sheltered slope. In Europe it is a plant of woodland and sheltered woodland edges, found in most places except the extreme north and south, and on the islands of the Mediterranean.

*Toxicology*
The berries contain xylostein but are of low toxicity, and poisoning is rare. Xylostein causes gastro-intestinal irritation and would produce symptoms of vomiting and diarrhoea.

# · HONEYSUCKLE ·

**_Lonicera periclymenum_ L.**

*Fruit*
The brilliant red berries of honeysuckle are a familiar sight in autumn hedgerows throughout Britain. The stalkless fruits are borne in dense clusters of five to fifteen, with a number of abortive berries between the ripe fruits. The globose berries are bright orange red, 8-9mm in diameter, with the pale brown, withered floral remnants prominent at the apex of the berry. The surface is glistening and sticky, so that it often becomes dulled with dust. The berries are very juicy, with orange coloured fleshy pulp. Most of them contain about six pointed oval, smooth yellow seeds, 4 × 2.5mm. They are ripe in August and September.

*Leaf and Flower*
Honeysuckle is a woody climber with stems up to 6m long which twine clockwise, often twisting round their own shoots to form 'cables' as they grow to seek support on other vegetation. The stems also run and root, and in dry woods they may carpet the ground, although in such places they rarely flower. Honeysuckle is deciduous, with lanceolate leaves which are subsessile, entire and opposite. They are never united at the base, as they are in the alien *Lonicera caprifolium*. The new leaves open in December and January from bluish-green buds, and the formed leaves may be slightly downy or hairless. The flowers appear from late May to late September, with two flowering peaks, in early June and September. The tubular flowers 40-50mm long are carried in simple umbels of six to fifteen, stalkless and with tiny bracts. The corolla is yellow with purple patches, and is two-lipped. The upper lip, the standard, is four-lobed, and the lower lip, the fall, is entire. There are five stamens and a long projecting style. The flowers are sweetly scented, especially at night, and the colour deepens to orange-buff after pollination. Honeysuckle is extremely variable in leaf and flower shape, and hybrids with other honeysuckles may occur. They tend to have long internodes on non-flowering shoots and are very robust, with tough, broad, leathery, almost connate leaves. They do not run or root on the ground. Some of them are shade-tolerant, and there they develop fat canes with reddish bark.

*Ecology and Distribution*
Honeysuckle is a common plant in woods and hedges throughout Britain, growing on a variety of soils and on rocks up to 600m. It is a typical climbing plant of oak woods, and is a major hedgerow plant. It avoids wet marshy areas and conurbations. It is widespread in west, central and southern Europe, extending north east to southern Sweden. The leaves are the feed-plant of the White Admiral butterfly (*Limenitis camilla*). The corolla tube appears to be too long for bees to pollinate, but the strongly scented flowers attract night-flying moths, especially hawk-moths, which act as pollinators.

*History and Uses*
The alternative name most commonly used for honeysuckle is 'Woodbine'. Culpeper advocated the use of the juice from the berries for treating eye conditions, and 'drink the juice thereof against the biting of an adder'. Gerard wrote: 'The floures steeped in oile, and set in the Sun, are good to annoint the body that is bemummed, and growne very cold.' A number of birds have been recorded eating the berries, especially blackbirds.

*Toxicology*
Poisoning with honeysuckle berries is rare, and the fruits appear to be of low toxicity. They contain valerianic acid and xylostein, which cause gastro-intestinal irritation and may produce vomiting and diarrhoea if a large quantity of the berries is consumed.

*Lonicera caprifolia*

This alien honeysuckle is distinguished by the connate upper leaves – hence its name of Perfoliate Honeysuckle. It is well established in the south of England from Kent and Sussex to Somerset, also in Cambridgeshire and Suffolk.

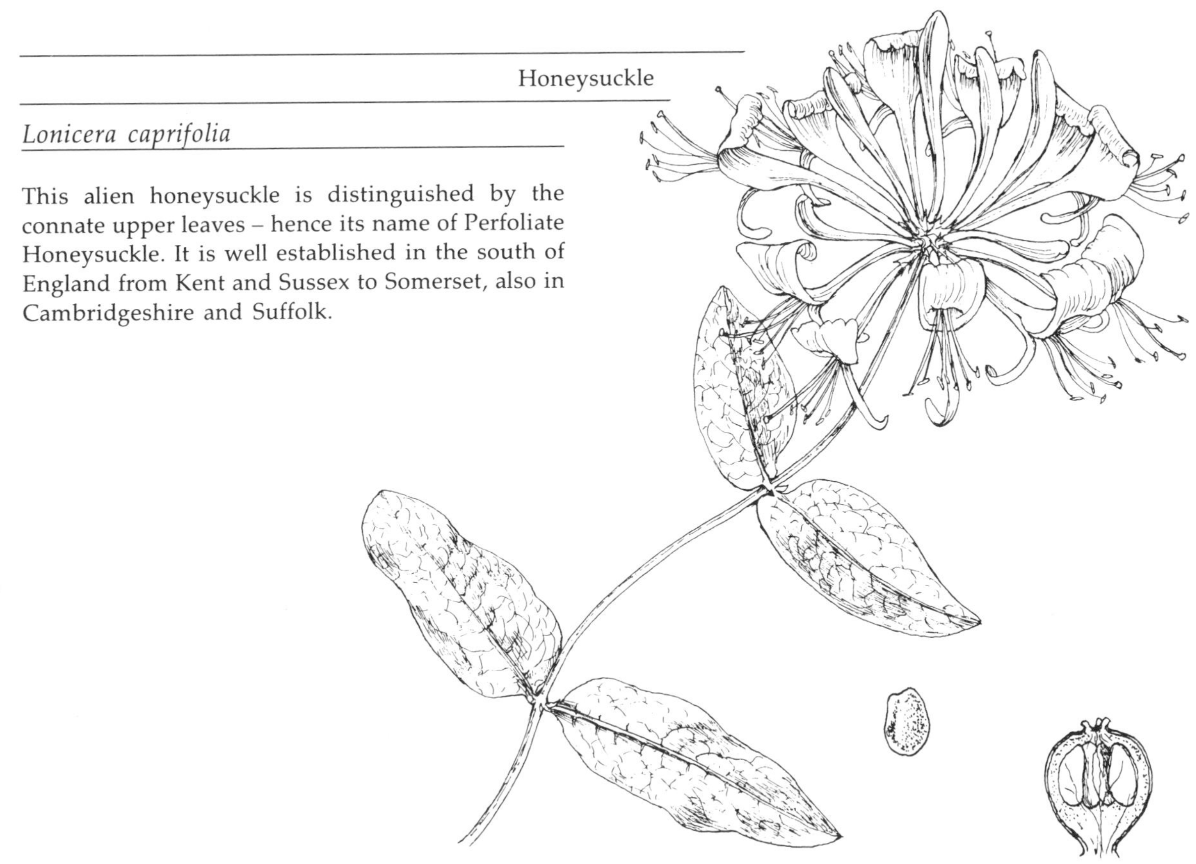

**Honeysuckle**

# · LILY-OF-THE-VALLEY ·

**Convallaria majalis L.**

POISONOUS

*Fruit*

The fruit of lily-of-the-valley is a flattened globose orange-red berry carried on a stout pedicel in a drooping, one-sided raceme, on a curving, ribbed scape. Fruit is not produced frequently, but five or six berries may ripen on a single plant. The berries are smooth and shining, 8mm in diameter, with soft flesh. They are divided into three cells, each of which may contain two seeds. Usually only one or two seeds mature, but fruits containing five full formed seeds have been recorded. The seeds are hard, pale brown, smooth and shiny, rounded triquetrous, 4.5 × 4mm, with a well-marked dark brown hilus. Ripe fruits may be found from August to November, by which time the foliage is shrivelled and brown.

*Flower and Leaf*

Lily-of-the-valley is a robust perennial with tough, spreading rhizomes, two or rarely up to four, broad, parallel-veined, lanceolate leaves with sheathed bases, and a leafless flowering scape 10-40cm tall. Flowers are produced from late May to mid-June, six to twelve in a drooping, one-sided raceme on a ribbed scape, each 5-8mm long, white and campanulate, with six recurved triangular lobes and a strong, very sweet scent. The bracts are narrow, membranous and shorter than the pedicels. There are six short stamens inserted halfway up the tube of the flower, and a single short style. Pink-flowered plants have been reported rarely – they were known from Tetton Woods in Somerset from 1872 to 1924, but have not been seen since.

*Ecology and Distribution*

Lily-of-the-valley is widespread on basic soils throughout England, and the north east and south east of Wales. It likes old undisturbed, dry woodland on limestone or sand, with good drainage, but an accumulation of leaf mould. Although it is widely planted it is also native, but decreasing throughout England. It occurs from Kent to Somerset, in the limestone woodlands of north Wiltshire and the Cotswolds in Gloucestershire, northwards to the rocky limestone woods of Yorkshire, Derbyshire and Cumbria; here it also grows in the sheltered grykes of the limestone pavement. It is very rare in central Scotland, being a likely introduction in Angus, and it is not native to Ireland. In Europe it grows widely in woods, scrub and mountain meadows in all areas except the extreme north and south.

*History and Uses*

Some of the sites where lily-of-the-valley grows have been known for centuries, such as the 'lily bed' of St Leonard's Forest in West Sussex. Legend relates that St Leonard fought with a dragon in the woods near Horsham. Where drops of his blood fell to earth lilies-of-the-valley sprang up.

Medicinal uses of lily-of-the-valley are long established. Gerard wrote: 'The floures of the Val-

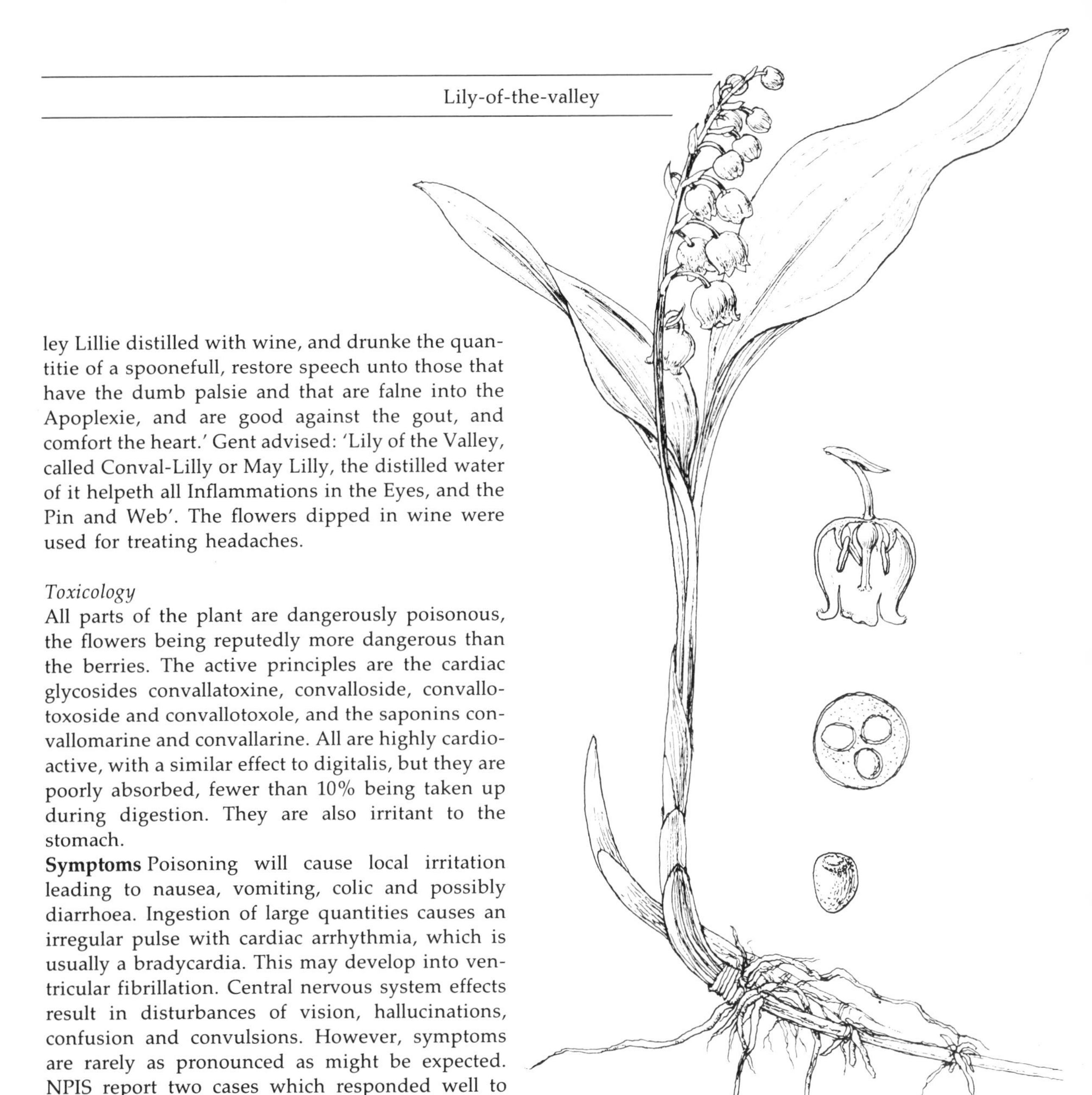

ley Lillie distilled with wine, and drunke the quantitie of a spoonefull, restore speech unto those that have the dumb palsie and that are falne into the Apoplexie, and are good against the gout, and comfort the heart.' Gent advised: 'Lily of the Valley, called Conval-Lilly or May Lilly, the distilled water of it helpeth all Inflammations in the Eyes, and the Pin and Web'. The flowers dipped in wine were used for treating headaches.

*Toxicology*
All parts of the plant are dangerously poisonous, the flowers being reputedly more dangerous than the berries. The active principles are the cardiac glycosides convallatoxine, convalloside, convallotoxoside and convallotoxole, and the saponins convallomarine and convallarine. All are highly cardioactive, with a similar effect to digitalis, but they are poorly absorbed, fewer than 10% being taken up during digestion. They are also irritant to the stomach.
**Symptoms** Poisoning will cause local irritation leading to nausea, vomiting, colic and possibly diarrhoea. Ingestion of large quantities causes an irregular pulse with cardiac arrhythmia, which is usually a bradycardia. This may develop into ventricular fibrillation. Central nervous system effects result in disturbances of vision, hallucinations, confusion and convulsions. However, symptoms are rarely as pronounced as might be expected. NPIS report two cases which responded well to simple treatment,and although older literature cites a few fatalities, more recent literature reports few symptomatic cases. Animal poisoning is rare, since access to the plant is uncommon, but small animals and birds have been poisoned. Chickens have been killed by eating the berries, and poisoning has also been reported in goslings.
**Treatment** Vomiting should be induced in children and gastric lavage used in adults, followed by the oral administration of 5-10 activated charcoal. Serum potassium should be checked, with ECG monitoring if necessary. Anti-arrhythmic drugs and cardiovascular support should be available if needed.

**Lily-of-the-valley**

# · MAY LILY ·

### *Maianthemum bifolium* (L.) F.W.Schmidt

*Fruit*

The fruit of May lily is a small, bright carmine red berry, 4-6mm in diameter, translucent-skinned and juicy. By the time that the berries are ripe in October and November the foliage has withered, so that the fruits, usually numbering five or six, are borne on the dry, withered stem on pedicels 3-4mm long. The berry is two-celled, with two ovules in each cell; most mature berries contain one to three seeds, the three-seeded berry having a rounded triangular shape. The seeds are flattened oval, 4 × 4mm, and white with a faintly pitted surface. The berries are soft and juicy, with a fruity smell and bitter taste. The juice is strongly pinkish-red staining. The flesh soon withers and may leave the exposed seed still attached to the pedicel. Fruit is seldom formed on plants in Britain.

*Leaf and Flower*

May lily is a perennial herb with a slender, creeping rhizome from which arises the annual stem 6-15cm tall. Non-flowering stems bear a single, long-stalked radicle leaf. Flowering stems have two cauline leaves, heart-shaped, sharply pointed and entire edged, with prominent parallel veins; the petioles sheath the main stem. The main flowering period is from early to late May, but some plants may still be found in flower in July. The flowers are erect, 2-4mm across, carried in dense racemes 10-40mm long, each flower subtended by a minute bract. The perianth has four free lobes, creamy-white and rounded, and four prominent exserted stamens, which give the flowering spike a frothy appearance. There is a single long stigma with two styles. The flowers are not scented.

*Ecology and Distribution*

May lily is a rare plant of deciduous woods on mainly acid soils where there is plenty of humus. It grows in Forge Valley Woods near Scarborough, Yorkshire, which may be its only true native site in Britain (there is an old record for 1860 near Thirsk),

and it is also recorded from Hunstanworth in Co Durham. Elsewhere, it is known in north Lincolnshire, in a wood near Swanton Novers in Norfolk where it may have been introduced with conifers, and near Dunwich Common in Suffolk, again a likely introduction. It was recorded from Ken Wood in Middlesex, where it was probably planted, in 1852 and as recently as 1925-26, and in south east Scotland it is established as an alien in the Lothians. In Europe it is widely recorded southwards to northern Spain, growing in humus rich soils, in woods and other shady places. On the island of Gotland in the Baltic it grows extensively and is becoming something of a pest, invading the margins of arable crops. Here the fruits do not colour up until October and November, when they are easy to spot against the first snowfall of the winter. Once, a late spring frost of −7°C in June resulted in few fruits being produced that year. The Swedish name is 'Ekorrbär' – squirrel-berry. The origin of the name is obscure.

*History and Uses*

May lily once occurred over vast areas of North America and Eurasia in the post-Glacial period, and it is known from as far east as Japan. Gerard wrote: 'The flowers of May Lillies put into a glasse and set in a hill of ants, close stopped for the space of a moneth, and then taken out, therein you shall finde a liquor that appeaseth the paine and griefe of the gout, being outwardly applied'.

*Toxicology*

The berries contain a cardiac glycoside, but there are no reports of poisoning in man or animals.

**May lily**

# · WHORLED SOLOMON'S-SEAL ·

***Polygonatum verticillatum*** **(L.)Allioni**

POISONOUS

*Fruit*

Whorled Solomon's-seal is without doubt the most elegant, as well as the rarest, of the three native species. The berries are carried in drooping clusters of two to four on peduncles 15-20 mm long arising from the axils of each whorl of leaves, each berry having a pedicel 6-10 mm long. The berry is divided into three loculi, each containing two seeds, of which only one usually develops to maturity. The immature berry is green, the cuticle marked with fine purplish dots, and the three developing loculi, divided by sulci, give the fruit an almost triangular shape. At this stage in August the papery remnants of the withered corolla often adhere to the fruit. The spots on the cuticle enlarge and coalesce as the fruit ripens and fills out, so by mid-September it is dark red with a purple suffusion, flattened globular, 6-7 mm in diameter, with three shallow grooves

down the sides and the sharp point of the stylar remnant at the apex. The flesh of the berry is dense, fairly hard, and usually contains a single yellow brown, flat ovoid seed, 5.5 × 4.5mm with a shiny, finely netted surface and a slight groove on either side.

*Leaf and Flower*

Whorled Solomon's-seal is a perennial with a stout rhizome and slender, erect, slightly angled annual stems 30-80cm high, hairless and often suffused with small purplish blotches. The leaves are linear-lanceolate, entire, with fine points, in sessile whorls of three to five up the stem. The flowers are borne on long peduncles in drooping axillary clusters of two to four. The perianth consists of six lobes united into a tube 6-10mm long, which narrows at the apex, white tipped with green, and unscented. The flowers are pollinated by bees or self-pollinated.

*Ecology and Distribution*

Whorled Solomon's-seal is now known only from a confined area on the banks of the tributary streams of the River Tay near Aberfeldy in Perthshire. It would appear to be extinct in its old site in Northumberland. It is a plant of rocky woods and scrub, growing under trees on the steep banks of the streams, where it is difficult to find. A flourishing and more accessible colony downstream from the Aberfeldy site had more than twenty fine flowering stems in 1984.

In Europe it grows in similar habitat from Arctic Norway to northern Spain, central Italy and south west Bulgaria.

*Toxicology*

There is no record of poisoning by this species in humans or animals, but it is wise to assume that it contains similar toxic principles to those found in the other two species of Solomon's-seal, and the consumption of berries or foliage should be avoided.

**Whorled Solomon's-seal**

# · SOLOMON'S-SEAL ·

***Polygonatum multiflorum*** **(L.)Allioni**

POISONOUS

*Fruit*

The fruit of Solomon's-seal is a smooth, globular berry, initially green, turning blue-black with a fine bloom on the surface when it is ripe in September. The berries are carried in drooping clusters of two to four on axillary peduncles about 25mm long, springing from the bases of the sessile, alternate leaves. The individual pedicels are 6-7mm long. The berries are 8-10mm in diameter, marked with three external grooves, hard in texture, and divided into three loculi. Each loculus contains two seeds almost equal in size, 5 × 5mm, irregular oval with a flat facet where they abut, smooth and buff-coloured. The apex of the berry bears the pointed stylar remnant.

Fruit is not produced in great quantity in most years by plants in Britain, and Salisbury reported a low set of viable seed. Those fruits which are produced often disappear from the plants before they ripen, possibly as a result of predation by slugs and snails, and certainly they have a very sweet, pleasant flavour.

*Leaf and Flower*

Solomon's-seal is a robust perennial with a stout rhizome, and round, arching annual stems 30-80cm long. The leaves are alternate in two rows, unstalked with the base sheathing the stem, ovate, pointed with an entire edge and many prominent veins. Flowers are produced from late May to early July in drooping axillary clusters of two to four, rarely as many as six. The flowers are tubular, 15 × 5mm, with a distinct concave waist to the tube, and six triangular recurved lobes to the perianth, often tinged with green. They are unscented and are self-pollinated or pollinated by bumble bees.

*Ecology and Distribution*

Solomon's-seal is a local plant of dry woods and hedgerows on basic soils, especially chalk and limestone. It is most frequent in southern England from Kent to Somerset, especially in Hampshire, Wiltshire and south Berkshire, where it is not uncommon. In South Wales it occurs in Montgomeryshire to Pembrokeshire and northwards to Radnorshire. In the north of England it occurs on

**Solomon's-seal**

limestone, especially in Cumbria. It is naturalized in Scotland and a few places in Ireland, but it is often confused with the garden hybrid *P.multiflorum* x *odoratum*, which is widely planted and naturalized. In Europe Solomon's-seal grows widely on calcareous soils in woodland and scrub, but is absent from parts of south west Europe and the extreme east.

*History and Uses*

The name may be derived from the rough Star-of-David pattern exhibited by a cut section of the rhizome, a design reputedly approved by King Solomon for putting evil spirits to flight. Other names are White Wort, White-root and Scala Coeli.

Gerard wrote: 'Galen saith, that neither herbe nor root here of is to be given inwardly, but the vulgar sort of people in Hampshire use bruised root in ale or white wine, or as an external poultice to knit fractured bones'. Gent noted: 'Solomon's-seal knitteth any Joynt, which by weakness useth to be often out of its place', and also 'Solomon's Seal is binding, the Roots of it is good in Wounds and Hurts, to cleanse them, and to dry and restrain Fluxes and Humors and bloody Flux and Lask; it is good for Ruptures and Burstness taken inwardly, or outwardly applied, and is good for inward and outward bruises'. Culpeper commented similarly, but had no use for the berries. The crushed rhizome has been used to remove freckles.

*Toxicology*
All parts of the plant appear to be poisonous. The seeds contain saponins and the glycoside asparagin, the rhizomes contain the aglycone diosgenin, and mucilages are also present. These are gastric irritant and cardiotoxic.
**Symptoms** Poisoning causes nausea, vomiting and diarrhoea, with blurring of vision, excitement and problems with cardiac conduction if very large quantities are consumed. NPIS have no cases on record, and there are few cases of human poisoning in the literature. Very few authenticated records of animal poisoning exist for the British Isles, but Baxter (1983) recorded poisoning in a 4½-month-old golden retriever, which ate a sufficiently large quantity of the leaves of the garden hybrid to pass recognisable plant material in its faeces. This puppy showed symptoms of a slowed heart rate and prolonged gastric emptying time, with uncontrollable vomiting. Exploratory laparotomy showed a low-grade gastro-enteritis, with congestion and oedema of visceral lymph nodes. At no stage was there any diarrhoea. No berries were present in the material consumed.
**Treatment** Vomiting should be induced if it has not already occurred. Fluid replacement therapy may be needed, and heart function should be monitored by ECG.

# · *ANGULAR SOLOMON'S-SEAL* ·

***Polygonatum odoratum* (Miller)Druce**

POISONOUS

*Fruit*
The fruit of angular Solomon's-seal is a drooping, rounded berry 8-11mm in diameter, sometimes slightly elongated, blue-black with a fine blue bloom on the skin when it is ripe in mid-September. The berry is usually solitary, rarely in groups of two or three, on fairly stout pedicels springing from the axils of the sessile leaves. It is marked externally by three grooves, indicating the divisions of the three loculi. Its skin is tough, as is the pithy content of the fruit. There are two ovules in each loculus, but usually only one or two seeds mature in each berry, the seeds being 4mm in diameter, roughly rounded with flattened facets and a broad scar at the base. The fruit has an unpleasant, foetid smell. It is not produced in any quantity in Britain, and although plants may flower well, by mid-August all the developing fruits will have gone – either predated or simply fallen off.

*Leaf and Flower*
Angular Solomon's-seal is a robust perennial, with a stout rhizome and tough, angled, slightly arching stems 15-30cm tall. The leaves are alternate and

**Angular Solomon's-seal**

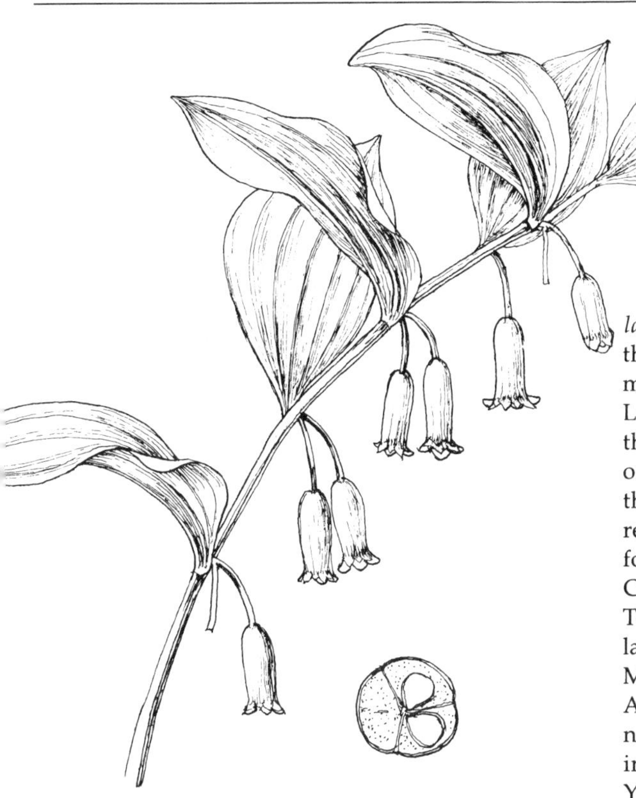

sessile, broadly ovate with a pointed tip, with many prominent veins and a midrib which runs down on to the slightly winged and twisted stem. The whole plant is much shorter and stouter than the other two Solomon's-seals, and the leaves lie close to the stem, with short internodes, so that they almost overlap. The drooping flowers appear from late May to early July, solitary or in groups of two or three on pedicels springing from the leaf axils. The tubular corolla, 18-22 mm long, has almost straight sides, and six triangular perianth lobes suffused with green, with wavy edges. The flowers are fragrant, which distinguishes them from those of the other two Solomon's-seals.

*Ecology and Distribution*

Angular Solomon's-seal is an uncommon plant of calcareous woodland and rocky limestone woods, occasionally of limestone pavement. In the limestone woodland of north Wiltshire and the Cotswolds in Gloucestershire it grows under a mixed canopy of ash and hazel, with the other specialities of old calcareous woodland, spurge-laurel (*Daphne laureola*), Herb-Paris (*Paris quadrifolia* and lily-of-the-valley (*Convallaria majalis*), favouring the more moist, sheltered slopes. It grows in similar sites in Leigh Woods near Bristol and the Cheddar Gorge in the Mendips, both in Somerset. It was lost from its only site in Hampshire when the A31 was widened through the New Forest, although a few plants were rescued and taken into cultivation. In Wales it is found at Wynd Cliff in Monmouthshire, Craig y Cilau in Breconshire, and in Pembrokeshire near Tenby and at Penally Burrows; a similar site to the latter, on calcareous sand near Harlech in Merioneth, is likely to be of introduced plants. Angular Solomon's-seal also grows on the carboniferous limestone of Staffordshire and Derbyshire, in calcareous woods on the Craven limestone of Yorkshire and Lancashire, and on the limestone of south Westmorland, although it has gone from its old sites in Cumberland. In Europe it grows widely in woods and rocky ground on calcareous soils, north to 66°N in Finland; on the alvar, the limestone pavement of the island of Gotland in the Baltic, it has been seen growing strongly in the sheltered cracks among the juniper bushes, in the parish of Stenkyrka above the famous Lummelunda caves. Pollination is similar to that in the common Solomon's-seal, either self-pollination or pollination by bumble bees. The foliage is often eaten to shreds by sawfly larvae.

*Toxicology*

The toxicology of angular Solomon's-seal is presumed to be similar to that of Solomon's-seal, the major active principle being convallamarin, which has a digitalis-like action upon the heart. Launert also described the presence of asparagin, mucilages and saponins. The berries are poisonous, and consumption of any part of the plant should be avoided.

**Symptoms/Treatment** As for Solomon's-seal.

# · HERB-PARIS ·

***Paris quadrifolia* L.**

POISONOUS

*Fruit*

The fruit of Herb-Paris is a solitary, four-celled fleshy capsule, borne upright at the tip of the annual stem, above the withering remnants of the symmetrical floral parts. The fruit resembles a dull, blue-black globose berry, 8-12mm in diameter, with the apex puckered in and crowned by the four withered styles. The sides of the fruit are marked by four grooves, which indicate the divisions into the four loculi. Each loculus contains two rows of seeds, up to eight in number, giving a maximum potential of thirty-two to each fruit, but normally only about eight seeds mature. The seeds are well rounded, 3 × 2mm, reddish brown and pitted. The fruit is ripe in July and August, but fruit production is sparse.

*Leaf and Flower*

Herb-Paris is a perennial with a thick, creeping rhizome and annual stems 15-40cm tall. There are usually four large stem leaves, pointed ovate and sessile, with prominent net veins, set symmetrically in a whorl halfway up the stem, although plants with three or as many as eight leaves have been recorded. The flowers are solitary at the top of the stem, 40-70mm in diameter, with eight pale green perianth segments in two whorls; the outer four are narrow lanceolate, and the inner four are thin and linear. The symmetry of the whole flower is a delight – each of the floral divisions laid out like a four-pointed star, one above the other, the sepals lined up with the divisions between the stem leaves and the petals set between the sepals at 45°. The eight erect stamens are attached to the bases of the perianth segments, with linear anthers, and there are four purple styles which crown the ripening fruit. The flowers appear from mid-May to mid-June, rarely as late as August.

*Ecology and Distribution*

Herb-Paris is a local plant of damp calcareous woodland. In England it is uncommon in north

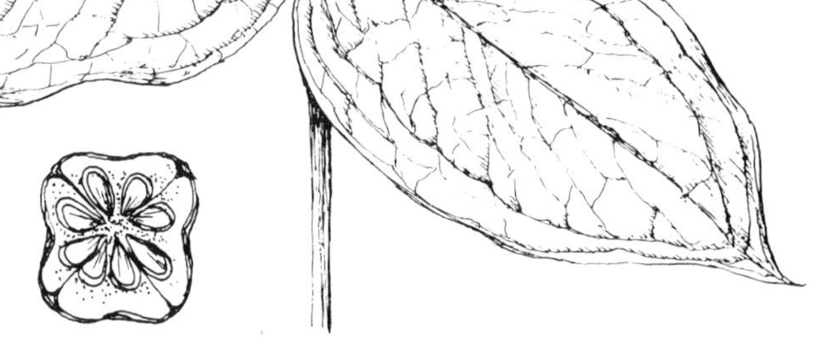

**Herb-Paris**

Kent and west Sussex in chalk woodlands and occurs west to the eastern half of Somerset in woods on limestone. In Suffolk it is widespread in most ancient woods and copses on the boulder clay, and it is similarly frequent in old woodland in north Wiltshire. In Wales it is found in the south east and east in Monmouthshire, Glamorganshire, Carmarthenshire, Brecon and Radnor, then northwards to Montgomeryshire, Flintshire, Denbighshire and Anglesey. In the north of England it is fairly common in limestone woods in Derbyshire, Yorkshire and Cumbria, becoming much rarer in Scotland, although it occurs as far north as Inverness. (There is an old record for 1895 from an islet in Loch Awe in Sutherland.) It is absent from Ireland. It is widespread in Europe in woods and damp shady places, but is rare near the Mediterranean. It tolerates fairly dense shade, but needs a moderate depth of humus and moisture to flourish. It often grows with dog's mercury (*Mercurialis perennis*), where, unless it is in flower, it is hard to see. The leaves of fruiting plants are withered by the time the fruits are ripe, making them equally difficult to spot. Most plants multiply by vegetative reproduction, forming localized colonies within a wood.

*History and Uses*

The first written mention of Herb-Paris is in 1548, under the name 'One-berry'. Gerard also knew it by that name, and as 'Herbe True-love'. He wrote: 'The berries of Herbe Paris given by the space of twenty daies, are excellent good against poyson . . . The same is ministered with great successe unto such as are become peevish'. Gent listed Herb-Paris berries under 'Simples that are good in general for Cattel that are bewitched. Two drams of the berries or seed of True-love, or one berry beaten to powder, and given him for twenty days together restoreth him'. He continued: 'Herb True-love, or One-berry is exceedingly cold, and the Leaves or Berries is good to expel Poison of all sorts, as also the Plague and Pestilence, is good for the Cholick, green Wounds, and to cleanse old and filthy Ulcers, and is good to discuss all Swellings in any part of the Body'. In the past the rhizome was used for its narcotic properties, to treat rheumatism and neuralgia.

*Toxicology*

All parts of the plant appear to be poisonous, especially the berries and seeds, but they are very unpalatable and so are seldom eaten. There are very few authenticated cases of poisoning. The active principles include a saponin-like substance paristyphnin, which on hydrolysis yields paradin – an active glycoside. This has a narcotic and irritant action.

**Symptoms** Poisoning causes nausea, vomiting, colic and diarrhoea, with localized irritation of the mouth. Ingestion of large quantities produces a dry mouth, dizziness, weakness, sweating and headache. There is difficulty in breathing, pain and difficulty in passing urine, and constricted pupils. Delirium and convulsions may follow. NPIS have no human cases on record, although a few cases of human poisoning and one fatality are described in the literature. Cases of poisoning in animals have been reported rarely in dogs and chickens, which showed an unsteady gait, abdominal pain and vomiting.

**Treatment** Vomiting should be induced if large quantities have been eaten. Kidney function and respiration should be monitored, and convulsions controlled with intravenous diazepam if necessary.

# · WILD ASPARAGUS ·

***Asparagus officinalis* L.subsp.*officinalis***

*Fruit*

The fruit of wild asparagus is a round, orange-red berry, 6-10mm in diameter, slightly flattened, carried singly or in pairs on pedicels 10-22mm long which are yellow and brittle by the time the berries are ripe in September and October. The dry, papery remnants of the flowers are often adherent to the bases of the berries, and there is a small pointed stylar remnant at the apex. The berries have a tough shiny skin and are quite juicy. They are divided into three sections containing one to six seeds, usually two or three. The seeds are 2-3.5mm in

diameter, flattened globular and black, with a finely pitted surface and a ventral radicle.

*Leaf and Flower*
Wild asparagus has a matted rootstock from which arise branched annual stems up to 2 m tall. The pale green, feathery foliage is composed of cladodes, leaf-like structures which are modified lateral shoots, 5-8 mm long by 0.3 mm wide, flexible and bristle-like, in whorls of five to seven. The dimensions of each whorl vary considerably, each having a small, brown, papery bract-like structure at the base of the whorl. The internodes between the whorls are normally 6-7 mm. The flowers appear from mid-June to August, the pale yellow, bell-shaped blooms borne singly or in pairs on pedicels 10-20 mm long. They are small, 4 mm across, with six recurved lobes, and they are usually dioecious, with an equal distribution of male and female plants. The female flowers are uniform, with one style and three stigmas, while the male flowers, with six stamens, show some intermediate sexual characters, and some are physiologically bisexual. Vazart maintains that certain plants could be truly monoecious. The flowers are insect pollinated.

*Ecology and Distribution*
Wild asparagus may not be a true native of the British Isles. It has been so widely cultivated for such a long time, and has escaped and established itself, especially in sandy soils near the sea and in waste ground with a light, well drained soil, that its possible status as a native plant cannot be determined. It is particularly common in south and south east England, from Kent and the Thames Valley west to Somerset, and frequently in East Anglia on the coast of Suffolk and in Cambridgeshire. It is not uncommon near the sea in Wales in sandy soils and waste places, becoming less common in the north of England, and it is absent from Scotland. It is very rare in Ireland. In Europe it is also widely cultivated and naturalized, our own stock probably being of European origin.

*History and Uses*
Asparagus has long been cultivated for its delicious young shoots which are a great delicacy. Culpeper used a decoction of the young branches for the treatment of 'Gravel and Stone out of the kidneys'. Asparagus has been used in the past for its diuretic and laxative properties.

*Toxicology*
The berries should be used with caution, and are probably best avoided. The young foliage can cause a contact dermatitis in humans.

**Wild asparagus**

# · SEA ASPARAGUS ·

***Asparagus officinalis* L.subsp.*prostratus* (Dumortier) Corbière**

*Fruit*
The fruit of sea asparagus is a bright, orange-red berry, approximately 8 mm in diameter, flattened globular, with a shiny skin and a small pointed stylar remnant at the apex. The berries are borne on drooping, jointed pedicels 4 mm long, often in among the densely bushy cladodes, and most are carried singly, although plentifully – up to thirty on a single frond. The six papery perianth remnants persist at the base of the ripening berry, which has a tough skin and a pulpy interior, with brightly orange-staining juice. The berries contain two to six dull black seeds, irregular oval or flat ovoid, 4.5 mm in length, with a minutely pitted surface. The berries are ripe in late September, after which they soon drop in the fierce winds which batter the coast in autumn.

*Leaf and Flower*
Sea asparagus is a perennial with a matted rootstock and branched annual stems up to 50 cm long, stiff and prostrate, with a markedly blue tinge to the foliage. The cladodes are short and stubby, 5-7 mm long and fleshy, in dense whorls of about nine, with much shorter internodes between the whorls than those of wild asparagus, so that the plants appear bushier and denser. In many plants the stems are only 10-15 cm long, growing straight out of the dry sand. The flowers are like those of wild asparagus, pale yellow green and campanulate, with six reflexed lobes, solitary on pedicels 2-6 mm long, and they appear to be truly dioecious. The flowers are scattered among the cladodes, unlike those of wild asparagus, and they flower earlier, in May and June.

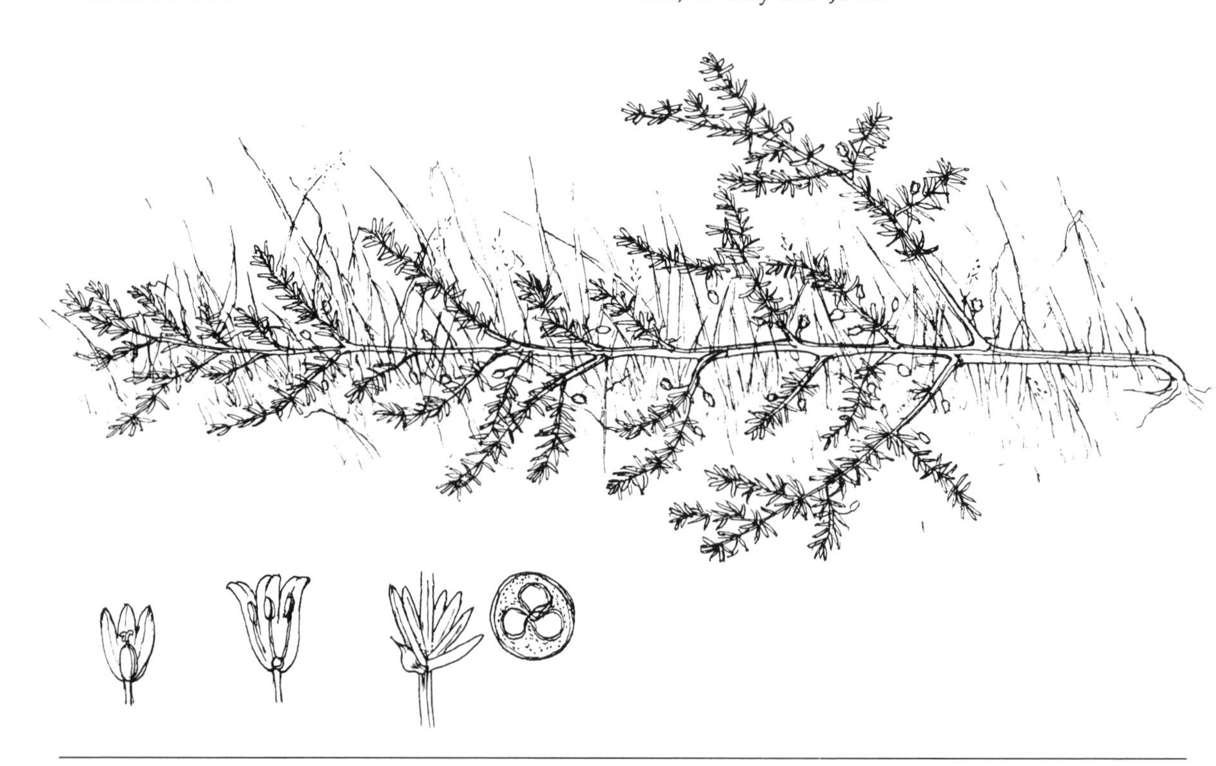

## Sea asparagus

*Ecology and Distribution*
Sea asparagus is a very rare native of sand dunes and sandy grass on sea cliffs, where it grows with such plants as thrift (*Armeria maritima*), wild carrot (*Daucus carota*), sea beet (*Beta vulgaris*) and rock samphire (*Crithmum maritimum*). By the time the fruits are ripe the surrounding grass is usually parched and brown, so that the lovely glaucous foliage and red berries create a vivid contrast. It likes a warm, sheltered slope with a predominately southerly aspect. It is now a rare plant, with a thriving colony on an island off the shore of the Lizard in Cornwall. It grows locally on the mainland, but has suffered from grazing pressure and drought, and from overgrowth by the alien *Carpobrotus edulis*. It was rediscovered on the coast of Pembrokeshire in the summer of 1982 on the grassy top of a limestone cliff, and it may still exist in other Welsh coastal sites in Glamorganshire, Carmarthenshire, Clwyd and Anglesey. In south east Ireland it grows on the coast in Wicklow and Waterford, and it flourishes in Wexford in the sand dunes of Ballysteige Burrows west of Kilmore Quay, among marram (*Ammophila arenaria*) and sea couch (*Agropyron pungens*). The older fruiting stems are often whitened at the base, and the plants are shorter and stubbier than the Cornish plants, probably as a result of the drier habitat. Elsewhere in Europe it grows on coastal rocks and maritime sands in the west, from northern Spain northwards to north west Germany.

*Toxicology*
There are no references to poisoning by this species, but as with wild asparagus the berries should not be eaten. Contact with the young foliage may cause dermatitis in humans. The foliage is palatable, and is certainly consumed by grazing animals with no apparent ill effect.

# · BUTCHER'S-BROOM ·

**Ruscus aculeatus L.**

*Fruit*
The fruit of butcher's-broom is a shiny red berry, 9-15 mm in diameter, an irregular, slightly flattened globe borne stemless in the centre of the leaf-like cladode. At the base of the berry are the six chaffy perianth remnants.

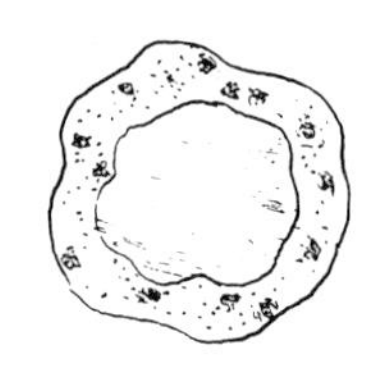

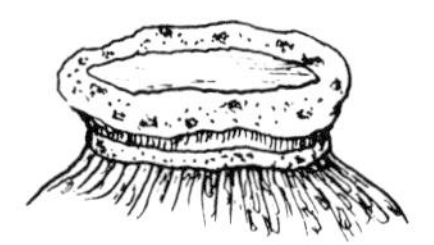

**Female flower**

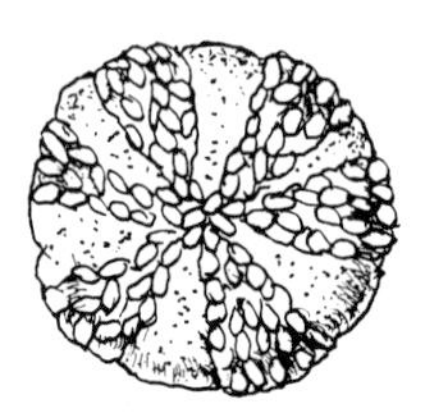

**Male flower**

The berry has a tough, leathery skin and contains one or two large seeds, rarely four. The seeds are rounded with a flat facet where they abut, approximately 6 mm in width, brownish yellow, smooth and shiny, with a large scar at the base. Ripe fruit may be found between October and May, with a peak in December and January, a welcome brightness set against the dark green foliage in a wintery wood. The berries take twelve months to mature.

*Leaf and Flower*
Butcher's-broom is a member of the *Liliaciae* and is the only shrubby monocotyledonous plant native to Britain. It is evergreen, with stiff, ribbed stems 40-120 cm tall, springing from a creeping rootstock, and dark green prickly foliage which has given it the apt name of 'Knee-holly'. The foliage is composed of cladodes, leaf-like modifications of lateral stems, which are ovate with a very sharp point. They are attached spirally with the base twisted through 90°, so that the blade lies in the same plane as the stem. Soft new shoots are produced from the rootstock in May and June, hardening rapidly to become indistinguishable from the older shoots. They are very vulnerable to frost. Flowers may be present at any time between October and May, and most plants show two flowering peaks, in November and December, and again in April. The flowers are like small stars, 3-5 mm in diameter, borne on very short stumpy pedicels in the centre of the cladode. The outer perianth segments are greenish white and pointed oval, with the three narrower, shorter inner segments set between them. The centre of the flower is occupied by a globular purple structure, the ovary of the female flower or the almost identical pseudo-ovary of the male, formed by the united stamen filaments.

Close examination with a powerful lens is needed to determine the sex of the flower. The female flower has a top shaped like a stumpy bottle, with a flat, translucent green stigma surface edged by a ring of granular cells blotched with purple. The top

**Butcher's-broom**

between the appearance of pollen and the closing up of the perianth, they are easily missed. In Guernsey Mrs Jennifer Page has found that andromonoecious plants (males with some hermaphrodite flowers), form 8.4% of the population and Dr Q. Kay has found it in 0.6% of plants in south Wales. Dr P. Yeo recorded one clone which produced male flowers for most of the winter, with a flush of hermaphrodite flowers in late spring, which rapidly developed fruit.

Butcher's-broom is a plant of dry woodlands, especially oakwoods and hedgerows. It reproduces vegetatively, so that unisexual clones may predominate in different parts of a wood, and some of the colonies are of great age, probably over one thousand years old. Confusion has arisen from plants appearing to bear male and female flowers

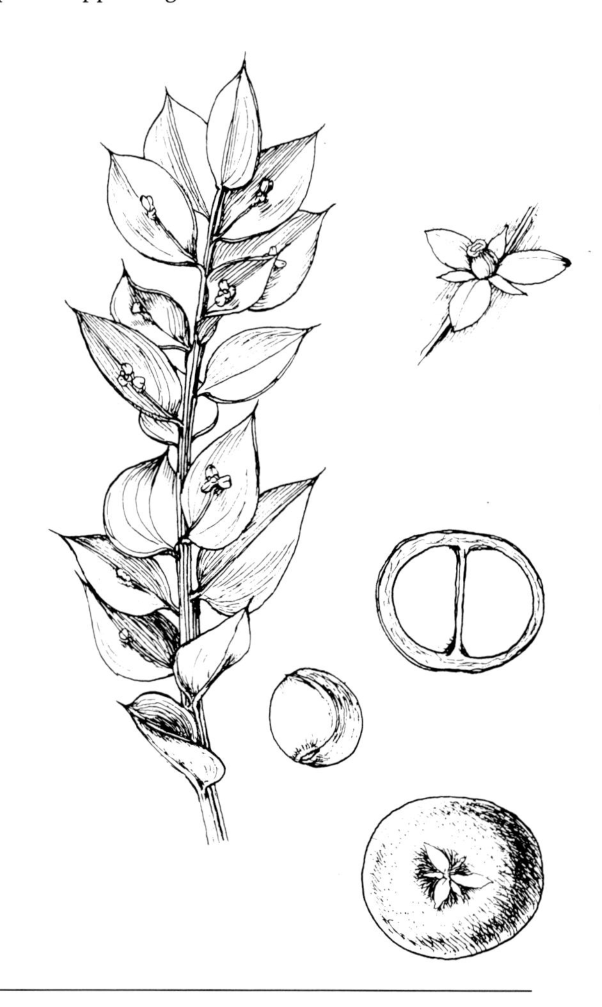

of the male flower is slightly more elongated and divided into six lobes, with golden coloured pollen grains borne in sections between the lobes.

*Ecology and Distribution*
Most plants are dioecious, producing either male or female flowers, but in certain areas plants may be monoecious, with male and female flowers on the same stem and rarely even on the same cladode. In addition, hermaphrodite flowers may occasionally be produced, but as only twelve hours may elapse

on separate shoots, but in many cases, when they are dug up and examined, they prove to be separate. Winter flowers are rarely fertilized, but insects are presumed to fertilize the spring flowers which produce fruit. The seven-spot ladybird *Coccinella septempunctata* hibernates on bushes near the shoot tips, and numerous spiders webs festoon them. Either species could act as pollinators.

Berry setting is variable, and most British plants carry few ripe fruits. Plants have been observed carrying a ripe red berry, a small green immature fruit and a male flower all on the same cladode in mid-October. Butcher's-broom is locally frequent in established broadleaf woodland and hedges in southern England, from Kent to Somerset, particularly in the Sussex Weald and the New Forest, in East Anglia where it is frequent in old woodland in south east Suffolk, and in South Wales from Monmouthshire to Pembrokeshire. It is also found in Caernarvonshire, Denbigh and Flint. Elsewhere it has been introduced. It is widely distributed in west, south and central Europe in scrub, woods and on sea cliffs, as far north as central England and Hungary.

*History and Uses*

Butcher's-broom was often planted in the past by apothecaries who used the root as a diuretic and for treating 'the stone'. The berries were used as a treatment for 'Knobbes' – from which Chaucer's Summoner suffered. According to the Doctrine of Plant Signatures, the berries on the cladodes resembled the red swellings and therefore could be used to treat them, hence the old name of Knobholm. Culpeper wrote: 'A poultice made of the berries and leaves being applied are effectual in knitting and consolidating broken bones or parts out of joint.' Gent similarly declared: 'A Decoction of the root of Butchers-broom or Knee Holly is good to conglutinate and knit things together, either inward or outward'. Also 'Butcher's-broom is hot in the second and dry in the first (*degree*) and is of a Cleansing Nature, it openeth Obstructions, provoketh Urine, expelleth Gravel and the Stone, and is good for the Strangury, Yellows and pain in the Head'. Deakin reported that 'The green branches tied together, and formed into brooms, were formerly used by butchers for cleaning their blocks', but this has been denied. Sometimes coloured berries were stuck on to the branches of butcher's-broom and the subsequent concoction was sold by pedlars as a Christmas decoration.

# · *BLACK BRYONY* ·

***Tamus communis* L.**

POISONOUS

*Fruit*

The fruit of black bryony is a rounded oval berry, 12mm long and 10mm in diameter, carried in short-stalked clusters in the leaf axils, unlike the berries of bittersweet (*Solanum dulcamara*), which are borne in stalked sprays. The fruits turn from green to yellow, and finally to a brilliant shining red when ripe from early September to December, hanging in dense garlands in the hedges after the leaves have withered and fallen. Each berry bears a sharply pointed remnant of the style at its apex, and is soft and juicy, containing one to six pale yellow, rounded seeds which are patterned with fine radial striations. The fruits are of unpleasant taste and produce a burning sensation in the mouth.

*Leaf and Flower*

Black bryony is perennial, with annual climbing stems. It is the only British member of the yam family, *Dioscoreaceae*, and derives its name from the colour of the underground tubers which are black, roughly ovoid, and may measure as much as 600mm in diameter, although a diameter of 200mm is more common. They lie deep in the soil, and from them arise slender but tough climbing stems, 2-4m in length, which twine clockwise around supporting plants. These stems die back below ground in winter. The leaves are heart-shaped, dark green and

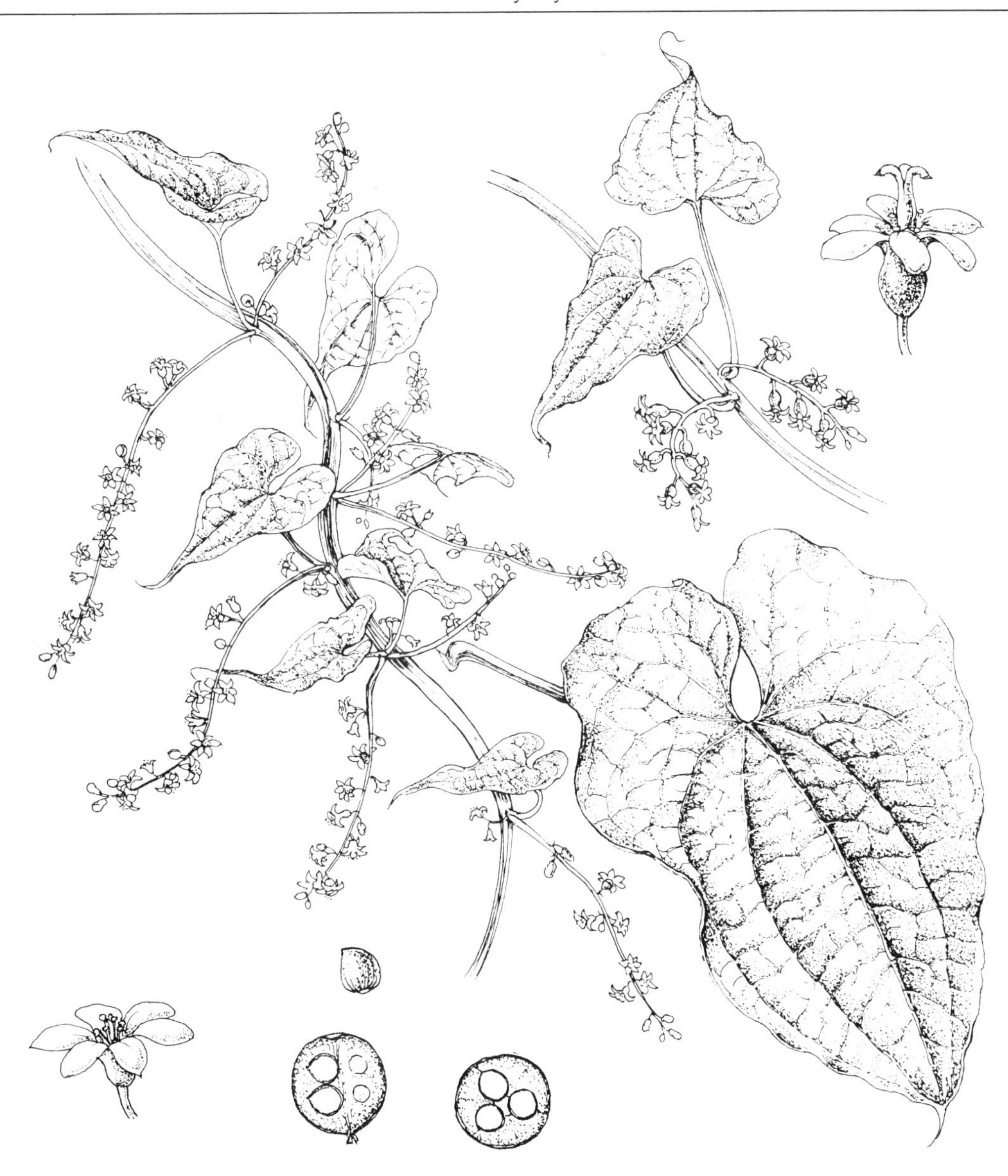

shiny with a pointed apex and deeply cleft base, and they have net veins. They are carried alternately up the stem on long stalks which may have two small stipule-like structures at the base, the stalks often twisting through 180° so that the leaf comes to lie on the opposite side of the main stem. The leaves are readily distinguished from the rough textured, palmate leaves of white bryony. Black bryony is dioecious, the male and female flowers appearing on different plants. The male flowers are borne in long, erect racemes springing from the leaf axils. The greenish-yellow flowers, 5mm in diameter, have six narrow campanulate perianth lobes, six stamens, and tiny bracts at the base of each flower stalk. The female flowers are almost sessile, in few-flowered, drooping clusters, each 4mm in diameter, with a single style and three two-lobed, recurved stigmas. The flowering period is from May to July, rarely as late as September.

*Ecology and Distribution*
Black bryony grows on wood margins, in scrub and hedgerows especially on moist, well drained, fertile soils and on calcareous land. It is common and widely distributed throughout England and Wales, as far north as the Lune and the Tees, but in all these areas it is absent from high moorland, mountains and coastal marshland. In Suffolk it has increased in frequency and is now recorded additionally from sandy and gravelly soils. It is absent from Scotland and the Isle of Man, and in Ireland it grows only around Lough Gill in Sligo and Leitrim, where it was probably introduced. It is likely that black bryony colonized Britain in the post-glacial period before the European landbridge was broken. In Europe it is common in the south, south central and western regions, north to Britain.

*History and Uses*
From the Middle Ages onwards the root of the black bryony was used as an emetic, and the fresh root or a tincture prepared from it was used to treat bruises. In view of the toxic nature of the plant, it well merited the old Suffolk name of 'Snakeberry'. The berries remain long in the hedges despite their prominent colour, because the toxic principle is distasteful, but this appears to decrease in amount in late winter, and some feeding by birds takes place.

*Toxicology*
The active principle is a narco-irritant resembling bryonin, the glycoside of white bryony, in its action, and it is concentrated mainly in the roots and berries. Other toxic factors include a histamine-like compound, calcium oxalate, and phenanthrine derivatives, with various mucilages and gums. The action of all these is highly irritant to the mouth and entire digestive tract.
**Symptoms** Poisoning produces a burning sensation in the lips, tongue and lining of the mouth, which may become swollen. There is nausea, vomiting and diarrhoea with abdominal pain. The eating of large quantities of berries causes paralysis of the respiratory muscles and death. NPIS have recorded thirteen cases without fatality, though references in literature suggest that fifteen berries may prove fatal to a child. There is some evidence that sheep and goats may eat the leaves without ill effect, but livestock can become addicted. Most poisoning cases occur when the tubers are exposed during ditch clearing. There are few recorded cases of poisoning in animals; three horses ate black bryony and died with symptoms of colic, a raised temperature and profuse sweating. In July 1985 a Jersey cow which had eaten the leaves of black bryony showed symptoms of severe ataxia and violent purgation.
**Treatment** Vomiting should be induced, especially in children if more than three berries have been eaten, and demulcents such as a mixture of eggs, milk and sugar, flour gruel or mixtures containing kaolin, should be given by mouth. The patient should be kept warm, and stimulants given if necessary. Should the pharynx be affected, care must be taken to ensure that the airway remains clear.

**Black bryony**

# · ITALIAN LORDS-AND-LADIES ·

**Arum italicum (Miller) ssp.neglectum (Townsend) Prime**

POISONOUS

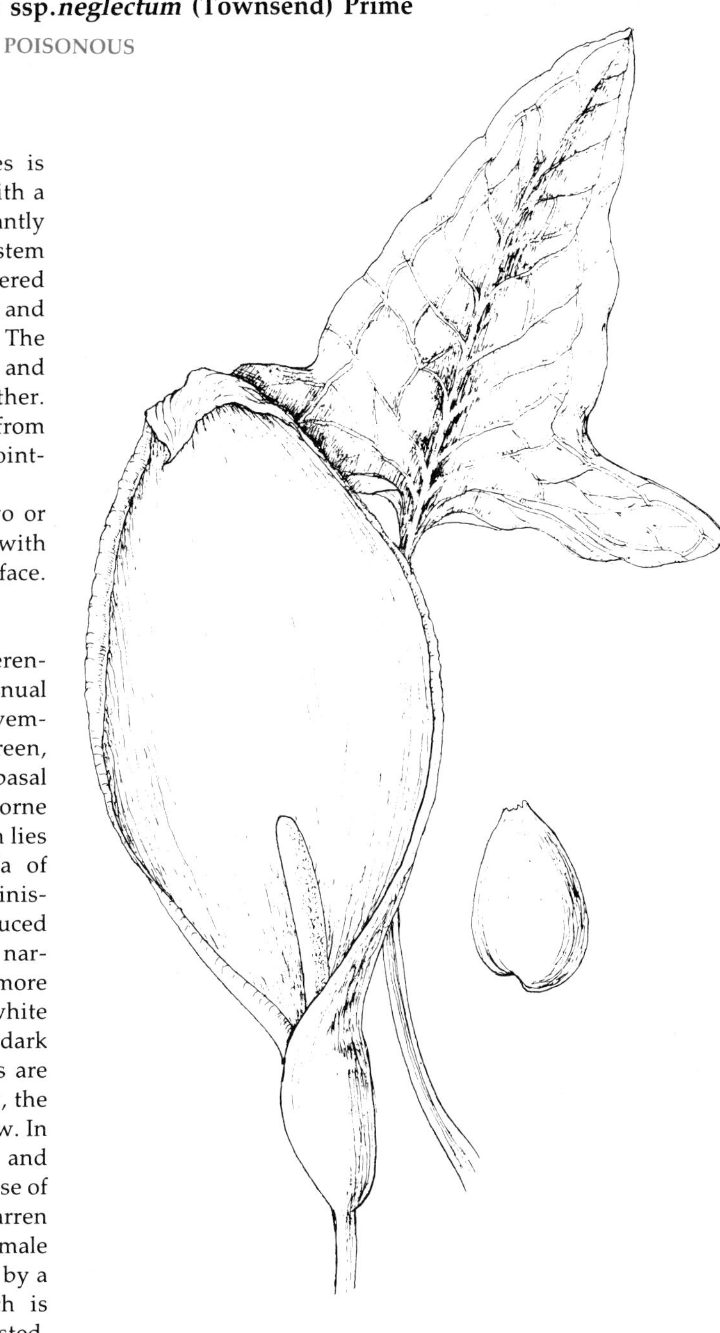

*Fruit*

The fruiting stem of Italian Lords-and-Ladies is very similar to that of the common species, with a cylindrical cluster of twenty to forty brilliantly shining orange-red berries on a stout, bare stem about 15cm tall, all the leaves having withered away by the time the fruits are ripe in August and September. They may persist into the winter. The berries are smooth and shining, 9-10mm long and irregular in shape where they lie against each other. The berries of *A.italicum* are quite distinct from those of *A.maculatum*, ending in a downward pointing beak at the outer pole.

They are soft and juicy, each containing two or three rounded, pale brown seeds, 4mm across, with one flattened or indented face and a netted surface.

*Leaf and Flower*

Italian Lords-and-Ladies is a stout-growing perennial with a large corm and four to eight annual leaves which first appear above ground in November. The hastate leaves are smooth and dark green, with a rounded apex and widely diverging basal lobes. The leaves are rarely spotted and are borne on long sheathing petioles. The leaf blade often lies parallel to the ground, creating an umbrella of foliage above the developing fruit head, reminiscent of a *Dracunculus*. The leaves of the introduced European species *A.italicum* ssp.*italicum* are narrower than those of the British native, with more widely diverging basal lobes and creamy white veins which create a striking contrast in the dark green leaf. The small male and female flowers are crowded in zones around the columnar spadix, the sterile upper part of which is smooth and yellow. In section the stem below the spadix is round and solid. The zone of female flowers is set at the base of the spadix, and there are zones of very long barren flowers above and below the upper zone of male flowers. The developing spadix is surrounded by a broad yellow-green, hood-like spathe, which is three times longer than the spadix, not waisted,

**Italian Lords-and-Ladies**

with the upper part of the spathe wide spread and the tip drooping over the spadix. Italian Lords-and-Ladies flowers from mid-May to late June, later than the commoner species, *A.maculatum*, after which the spathe rapidly withers.

*Ecology and Distribution*

Italian Lords-and-Ladies is a native British species and is considered by Professor Malvesin-Fabre to have existed in Britain from the mild Atlantic period about 6000 BC. It appears to like some overhead shelter and a deep, well drained soil. The site must not be subject to frost, since the early emerging leaves would be damaged, and it does not grow in soils with a pH lower than 5.5. The absence of persistent snow in winter is an important factor, and a generous water supply is more important than light. At one site in Sussex the plants occupy a distinct zone at the damp foot of a steep slope, with *A.maculatum* abundant in the drier upper zone. Pollination is effected by midges of the genus *Psychoda* and probably some flies. In Britain it is distributed along the south coast from west Sussex to Cornwall, with its furthest site inland a few miles north of Selborne in Hampshire, and an outlying centre in a sheltered limestone gorge in Glamorganshire. Apart from sites near Arundel in West Sussex and in east Hampshire, it grows on the east side of the Isle of Wight, where it was first recorded by A Hambrough in 1854 near Ventnor. It grows on cliffs south west of Swanage in Dorset, and is abundant in the south of Portland. It grows in Devon in sheltered coastal lanes, and on cliffs in Cornwall from Polperro around the Lizard to Newquay on the north coast. Both this subspecies and ssp.*italicum* have been widely cultivated and introduced, not only in southern England, but also in Wales, the Midlands and north to Kirkudbrightshire. In Hampshire Dr Prime recorded ssp.*italicum* growing in the middle of colonies of ssp.*neglectum*, and in Portland he found both subspecies and intermediate forms. Both subspecies are widespread in western Europe and the western Mediterranean: they grow in lush abundance on Menorca in the Baleares. The range extends eastwards to Crete and Cyprus. Many flowering spikes are picked off by birds while still in bud, and they are frequently chewed by snails and slugs. Ripe fruits are taken by birds, which probably act as seed dispersal agents.

*Toxicology*

There are no records of poisoning with this species in humans or animals in Britain, but it would appear to be in all respects as unpleasant and dangerously toxic as common Lords-and-Ladies.

# · LORDS-AND-LADIES ·

***Arum maculatum*** **L.**

POISONOUS

*Fruit*

The berries of Lords-and-Ladies are carried in a dense cylindrical cluster on top of a stout stem 20-30 cm high. Most plants bear twenty to thirty berries, each 10-15 mm long, shiny orange-red, and slightly irregular in shape, with the apex puckered in and lacking the beak typical of *A.italicum*. The flesh of the berry is orange and contains a nauseous-smelling juice with an acrid taste. Each berry contains seven ovules arranged in two rows, but over 80% of ripe fruits only contain a single pale brown globular seed, 4.5 mm in diameter, with a netted surface. The berries are ripe from July onwards, by which time the spathe and leaves have withered away, and some fruits persist until December.

*Leaf and Flower*

Lords-and-Ladies is a stout perennial, growing from a scaly corm, with seven to twenty arrow-shaped leaves up to 40 cm long, each with a long sheathing base to the petiole. The first leaves appear in January and February, and show considerable variation in their size, shape and lustre. Many plants have leaves spotted with purple anthocyanin, which gives the plant its specific name *maculatum*. The degree of spotting is variable; colonies in the north of England show a progressive decrease in spotting, until in Scotland, at the edge of its range, it is a rare characteristic. The small unisexual flowers which have no sepals or petals are borne in zones on a columnar structure 70-120 mm long called the spadix. In the developing stage the spadix is enclosed by the hood-like spathe, which is narrower than that of *A.italicum*, and has a pronounced waist at the base. The spathe shows a wide range of colouring with anthocyanin, from pale yellow-green to dull purple, uniformly or patchily coloured, or fringed with purple. At the base of the spadix is a zone of female flowers, above which is a zone of sterile ovaries which do not develop ovules. The male flowers in the next zone up each bear two protruding stamens, and the upper smooth part of the spadix is sterile, usually purple-coloured but occasionally butter-yellow. In section the stem below the spadix is oval in shape, and solid. The flowers appear from mid-April to late May. The process by which pollination occurs is curious and complex: the spathe unfolds in the late morning or early afternoon, and appears attractive to moth-flies such as *Psychoda phalaenoides*. The temperature of the upper end of the spadix rises as the plant uses up stored starch grains for energy, until by late afternoon it may become 15°c higher than that of the surrounding air, and quite warm to the touch. It also develops a foetid smell. Large numbers of insects are attracted, and on entering the cavity of the spathe are trapped by hairs near the entrance. Up to 4,000 have been recorded in a single spathe, mainly female *Psychoda* sp. At this time the stigmas are receptive to pollen, but by the next morning they have withered and the spathe has cooled. The midges feed on available nectar and the stamens mature, shedding copious pollen over the trapped insects. Finally the trap hairs shrink allowing the insects to escape. Self-pollination does not occur. By the end of July the leaves have gone and the spathe shrivelled to a brown tissue-paper remnant around the developing berries.

*Ecology and Distribution*

Lords-and-Ladies is a very common plant of woods and hedges and shady places throughout southern England and Wales. It prefers a base rich soil and does not grow in peaty acid soils. It is less common in northern England and Ireland, and reaches its northern limit of distribution near the Firth of Forth. The northern plants tend to be smaller. It is found up to 330 m in Derbyshire and 400 m on Lock Fell in Lancashire, but it is very frost-sensitive because the leaves open so early in the year. In cold weather the spathes do not open, and the pollinating insects are inactive. It is absent from the Highlands, Hebrides, Orkney and Shetland. *A.maculatum* is widely distributed across the central part of western Europe as far east as the Black Sea. Slugs and snails devour the ripe berries, but birds are the main agents of seed dispersal. Blackbirds, thrushes, pigeons and pheasants have all been recorded eating the fruits. Small mammals dislike the pulp of the berries and avoid eating the seeds.

*History and Uses*
Geoffrey Grigson in *The Englishman's Flora* (1955) noted that over one hundred names have been recorded for *A.maculatum.* The name Lords-and-Ladies does not appear before the 17th Century and may derive from the purple or yellow spadix. The earliest English name, 'Cuckoo-pint', has an Anglo Saxon origin from the words cucu (lively) and pintle (pint or penis), and it was held to have aphrodisiac properties. Other common names were Jack-in-the-pulpit, Wake-robin, Aron, Calf's-foot and Rampe. Culpeper wrote: 'The juice of the berries . . . provoketh urine and bringeth down women's courses, and purgeth them effectually after childbearing, to bring away the after-birth. Taken with sheep's milk it healeth the inward ulcers of the bowels . . . The juice of the berries boiled in oil of roses . . . and dropped into the ears, easeth pains in them.' In 1681 Gent wrote: 'Cuckow-Point or Wake-Robin is hot and dry in the third degree, it is good given against the Plague or Poison, being mixed with Vinegar, it is good for short-windedness and Cough of the Lungs, it is good to provoke Urine, is good for the Itch, Ulcers, and to take away the Pin and Web in the Eye.' In the reign of Elizabeth I, starching of ruffs and linen became very popular, and in 1597 Gerard wrote: 'The most pure and white starch is made of the roots of Cuckowpint; but most hurtfull to the hands of the Laundresse that hath the handling of it, for it choppeth, blistereth and maketh hands rough and rugged and withall smarting'. This effect was due to the crystals of calcium oxalate which it contained. Deakin wrote: 'In the island of Portland and about Weymouth, where the plant grows in great abundance, the roots are prepared by maceration, powdering and washing away the juice; the residue is then a wholesome and nutritive article of food,

**Lords-and-Ladies**

sold under the name of Portland Sago.' This was certainly sold in 1797 by a Mrs Jane Gibbs of the Portland Arms, Portland, but by 1875 the practice had ceased. An infusion of the leaves was used for treating coughs, and lung and stomach ailments.

*Toxicology*
All parts of the plant are highly toxic, especially the berries, containing a coniine-like substance, the saponin arin which haemolyses red blood cells, and cyanogenic glycosides. The effects of poisoning are produced very rapidly, often within thirty seconds.
**Symptoms** In man there is severe inflammation of mucous surfaces, especially the lips and mouth, with an intense burning sensation, swelling of the lips and tongue, and possible occlusion of the airway. There is vomiting and violent enteritis, tachycardia, and in some cases the development of a rash of red spots over the entire body surface. Despite the visual attractiveness of the berries, the rapidity with which symptoms arise prevents large quantities being consumed by children, and cases of poisoning are uncommon. O'Moore investigated six outbreaks of poisoning in cattle. Symptoms were not consistent, but included profuse salivation, oedema of the neck, incoordination, convulsions and death. At postmortem examination the entire alimentary tract and gall bladder were inflamed. Kyle recently recorded poisoning in a thirteen year-old Welsh greyface ewe which exhibited intermittent muscle weakness and profuse green diarrhoea with a sour smell. Her suckling lamb showed depression. The ewe recovered spontaneously in four days, and then went straight back to the Lords-and-Ladies and started eating it again.
**Treatment** Vomiting should be induced promptly to clear any berries remaining in the stomach, and demulcents such as eggs, milk, sugar and oatmeal or starch given orally. Care must be taken to ensure that the airway remains clear, remembering that the mucosal surfaces of the mouth and pharynx will be painfully inflamed.

# · GLOSSARY ·

*Accrescent* Enlarging after flowering
*Achene* A dry fruit containing one seed and not splitting open.
*Adventitious* Applying to a root or shoot produced in an abnormal part of a plant.
*Aggregate* A group of closely related species treated as a single species.
*Aglycone* The active, non-sugar, part of a glycoside.
*Alder carr* Wet, fen-like areas with a strong growth of alder or willow trees.
*Alien* A plant introduced by man.
*Alkaloid* A complex organic substance, often poisonous.
*Andromonoecious* Having male and hermaphrodite flowers on the same plant.
*Anther* The part of the stamen bearing the pollen grains.
*Anthocyanin* Red, purple and blue plant pigments.
*Apical* At the tip (of a shoot).
*Apomictically* Produced by apomixis.
*Apomixis* Reproduction by seed not formed as a result of sexual fusion.
*Appressed* Closely applied to, but not joined to, another structure.
*Arboreal* Associated with trees.
*Aril* Fleshy cup surrounding the seed of yew and several other plants.
*Arrhythmic* Applied to a disturbed heart rhythm.
*Arteriosclerosis* Hardening of the wall of arterial blood vessels.
*Asexual* Not related to normal sexual reproduction.
*Astringent* Having a drying effect upon the mouth or gut.
*Ataxia* Condition involving loss of balance and normal ability to walk.
*Atropinesterase* Enzyme which inactivates atropine.
*Autonomic* The nervous system not under conscious control.
*Axil* The angle between the stem and a leaf stalk.
*Berry* Fleshy fruit formed from ovary wall, containing one or more seeds which lack a stony covering.
*Biennial* Plant taking two years to complete its life cycle.
*Bract* Modified leaf at the base of a flower stalk with a flower in the axil.
*Bracteole* Small leaf on a flower stalk lacking a flower in the axil.
*Calcareous* Containing chalk or lime.
*Calcicole* A plant which enjoys calcareous conditions.
*Calcifuge* A plant which avoids calcareous conditions.
*Calyx* The whorl of sepals, sometimes fused into a tube.
*Campanulate* Bell-shaped.
Capsule A dry structure containing several seeds, splitting open to release them.
*Carboniferous* Pertaining to a system of rocks formed between the Devonian and Permian periods, sometimes containing coal.
*Carpel* One of the divisions of the female part of the flower.
*Cartilagenous* Containing cartilage or having a similar tough texture.
*Cauline* Leaf carried on the aerial part of a stem but lacking a flower or inflorescence in the axil.
*Chromosomes* Deeply staining structures within the cell nucleus which carry the genes responsible for inherited characteristics.
*Cladode* Modified stem flattened to resemble a leaf.
*Clone* Group of plants which have a precisely similar genetical make-up because they have been produced vegetatively or apomictically.
*Coniferous* Bearing cones (e.g. pine trees).
*Connate* Structures of the same kind which become joined together, although distinct in origin.
*Corolla* The whorl of petals, sometimes fused into a tube.
*Corrosive* Having a destructive effect upon tissues.

*Corymb* A raceme with pedicels of varying length, so that the flower head is flat-topped.
*Cotyledon* The first leaf or leaves of a plant after germination of the seed.
*Coumarin* Descriptive of a group of glycosides.
*Cultivar* A plant form produced by horticulture.
*Cuneiform* Wedge-shaped, with a pointed base.
*Cuticle* Outer cell layer of a plant structure, often waxy.
*Cyanogenic* Compound yielding hydrocyanic acid.
*Cyme* A forked inflorescence where each terminal point bears a flower.
*Cystitis* Inflammation of the urinary bladder.
*Deciduous* Shedding leaves annually in autumn.
*Deflexed* Bent sharply downwards.
*Demulcent* A soothing medicine for internal use.
*Depauperate* A growth-form of less than normal size.
*Dermatitis* Inflammation of the skin.
*Dichotomous* Branching equally into two parts.
*Dioecious* Having male and female flowers on separate plants.
*Diploid* Having two basic sets of chromosomes.
*Disjunct* Applied to populations of plants or animals which are widely separated.
*Diuretic* Stimulating urine production.
*Drupe* Fleshy fruit formed from the ovary wall, usually containing a single seed surrounded by a hard, stony case.
*Drupelet* A small drupe, often part of a compound fruit.
*Eglandular* Without glands.
*Ellipsoid* A solid object, elliptical in longitudinal section.
*Embryosac* The cavity in the ovule within which the embryo plant is produced.
*Epiphytic* Growing upon another plant, but not parasitic on it.
*Filament* The stalk of the anther.
*Flaccid* Soft and floppy, lacking in cell tone.
*Florula* The plant species of a small geographical area.
*Flush* A patch of ground, usually on a slope, where water flows diffusely.
*Garigue* Mediterranean vegetation composed of low bushy and spiny shrubs interspersed with open patches of bare ground.
*Gastro-intestinal* Relating to the stomach and intestine.
*Glabrous* Smooth and hairless.
*Glaucous* Bluish.
*Globose* Ball-shaped.
*Haematuria* The presence of blood in the urine.
*Hastate* Shaped like a spearhead.
*Haw* The fruit of the hawthorn, a pome.
*Hepatitis* Inflammation of the liver.
*Herbaceous* Having a soft texture like a leaf.
*Hermaphrodite* Flower containing male and female organs.
*Hilus* The scar left on a seed by the stalk of the ovule.
*Hip* The fruit of the rose.
*Hybrid* Plant originating by the fertilization of one species by another.
*Hydrocyanic acid* Poisonous substance produced in the hydrolysis of cyanogenic glycosides.
*Hydrolysis* The chemical interaction of a substance with the hydroxyl (-OH) ion of water.
*Hyperaesthesia* A pathological increase in sensitivity to stimuli.
*Inflorescence* The flowering portion of a stem above the last stem leaves.
*Internode* Part of the stem between two adjacent nodes or leaf joints.
*Introgression* The process of repeated back crossing of a first generation hybrid with one of its parents.
*Involucre* Whorl of bracts forming a collar below a compound flower head.
*Lanceolate* Shaped like a lance head – tapering and pointed.
*Laparatomy* Operation to open the abdomen.
*Leached* Pertaining to soil from which the soluble elements have been removed by the action of water.
*Leishmaniasis* A disease, mainly of tropical regions, caused by infection with bacteria of the genus *Leishmania*.
*Lenticel* Small, scale-like structure, shaped like a lentil.
*Loculicidal* Splitting down the middle of each cell of the ovary, often relating to capsules.
*Loculus* Small, separate chamber or cell.
*Magnesian* Rock system, containing the mineral magnesia, formed in the Lower Permian period.
*Maquis* A Mediterranean vegetation type composed of tall shrubs and scattered trees.
*Meiosis* The process of division of a cell nucleus during sexual reproduction.
*Melanic* Containing dark pigments such as melanin.

*Melanin* Dark, almost black, pigment present as granules in cells.
*Mesolithic* Geological period, the Middle Stone Age.
*Monocotyledon* Plant having a single cotyledon (grasses and lilies).
*Monoecious* Having separate male and female flowers on the same plant.
*Monosaccharide* Simple sugar such as fructose.
*Montane* Relating to mountain habitat.
*Mucilagenous* Containing mucilage – a slimy, viscous substance.
*Native* Relating to plants already in a country, not resulting from human introduction.
*Neolithic* Geological period, the Late Stone Age.
*Nutlet* A small dry seed with a hard wall.
*Oedema* Pathological condition of excess serous fluid within tissues.
*Opisthotonus* Spasm of the body, in which the spine is arched backwards or upwards.
*Orbicular* Rounded, as a leaf of equal length and breadth.
*Ovary* The lower part of the female reproductive organ, containing the seeds, composed of one or more carpels.
*Ovule* Structure containing a single egg, developing into a seed after fertilization.
*Oxalate* A salt formed of a base combined with oxalic acid.
*Palmate* Divided like the fingers of a hand.
*Pedicel* The stalk of a single flower.
*Peduncle* The stalk of an inflorescence or part of it.
*Peltate* A flat organ, such as a leaf, with the stalk attached underneath, not at the edge.
*Pentaploid* Having five basic sets of chromosomes.
*Perennial* A plant living for more than two years.
*Perianth* The outer, non-reproductive, parts of a flower, divided into the inner petals (corolla) and outer sepals (calyx), sometimes fused into a tube.
*Perigynous zone* The annular zone between the attachment of the female organs and the outer male organs and perianth.
*Perirenal* Around the kidney.
*Peristalsis* The wave-like contractions which propel food through the bowel.
*Petaloid* Resembling a petal.
*Petiole* The stalk of a leaf.
*pH* A logarithmic measure of the hydrogen ion concentration in moles per litre. Values greater than seven are alkaline, less than seven are acidic.
*Photosensitization* The pathological process rendering a tissue damagingly sensitive to light.
*Pinnate* Of a compound leaf, the leaflets arranged in opposite pairs on either side of a midrib.
*Pinnatifid* Pinnately cut, the divisions not reaching the midrib.
*Pollarded* (Of a tree) cut to within 2-3m of the ground to produce a crop of thin branches for fencing or withies.
*Polypeptide* A protein-like substance formed by a chain of aminoacids.
*Polyploid* Having more than two basic sets of chromosomes.
*Pome* The fruit of apples, pears, etc. A succulent false fruit formed by the receptacle wall, containing seeds with a tough, not stony, coat.
*Procumbent* Lying loosely on the surface of the ground.
*Pruinose* Having a grape-like bloom on the surface.
*Pseudo-ovary* A structure in a male flower having an external resemblance to a female ovary.
*Pubescence* A soft, short, hairy coating.
*Pyrene* The stone of a fruit such as a drupe.
*Pyriform* Pear-shaped.
*Raceme* An unbranched flower-spike where the flowers are borne on pedicels.
*Radicle* The part of an embryonic plant which forms the root.
*Receptacle* The tip of the flower stalk to which the sepals, petals and other structures are attached.
*Refractile* Having the quality of refracting light, thus contrasting with surrounding material.
*Rhizome* An underground stem, usually growing horizontally and lasting more than one season.
*Rhodologist* An expert in the genus *Rosa*.
*Ruminant* An animal with a complex stomach, including the rumen, such as a cow or sheep.
*Ruminotomy* The operation to open and explore the rumen.
*Saponin* A glycoside which causes a frothing effect upon water.
*Scape* The flowering stem of a plant of which all the foliage leaves are attached basally.
*Schistose* Rock having a crystalline, layered texture.
*Sclerotic* Having a hard, stony texture.
*Sepal* A division of the calyx, one of the outer whorl of perianth segments.
*Serrate* Toothed like a saw.

*Sessile* Stalkless.
*Silicaceous* Containing hard granules of silica.
*Sinuate* Having a wavy edge or margin.
*Solanaceous* Belonging to the family *Solanaceae* (eg. bittersweet).
*Spadix* A dense, erect spike of florets found in flowers of the Arum family.
*Spasmolytic* Referring to a drug which will alleviate spasm of (smooth) muscle.
*Spathe* A large bract or pair of bracts which envelop a developing flower spike.
*Stigma* The receptive upper part of the female reproductive organ.
*Stipule* A leaf-like appendage at the base of a leaf stalk, often paired.
*Stolon* An above-ground, creeping stem.
*Stone* The seed of a drupe, surrounded by a hard coat.
*Striated* Marked with fine longitudinal furrows or ridges.
*Strobilus* A fruit like a pine cone, hence the aggregated male flowers of yew, shaped like a tiny pine cone.
*Style* The stalk arising from the tip of the ovary which bears the stigma(s) at its tip.
*Sub-erect* Not quite upright.
*Sub-globose* Having a shape not quite spherical.
*Sub-maritime* Relating to plants growing near the sea but not on the shoreline.
*Sub-orbicular* Of a leaf having a shape not perfectly circular.
*Sulcus* A deep groove.
*Taxa* Division in plant classification such as a genus or species.
*Tendril* A curling, climbing organ often formed by modification of the terminal leaflet of a pinnate leaf.
*Tetraploid* Having four basic sets of chromosomes.
*Tomentose* Densely covered with fine hairs.
*Toxic* Poisonous.
*Toxin* Substance having a toxic character.
*Trefoil* Divided into three leaflets.
*Trifoliate* Divided into three leaves or leaflets.
*Triquetrous* A solid structure triangular in section and sharply angled.
*Tuber* Swollen part of a root or underground stem used as a storage organ.
*Umbel* Umbrella-shaped flower-head with branches arising at the same level, often flat-topped.
*Understorey* Plants forming a community below the main level of tree growth.
*Urethritis* Inflammation of the urethra, the exit canal from the urinary bladder.
*Vasodilator* Substance acting upon the muscle coat of blood vessels to increase the internal diameter of the vessels.
*Vermifuge* Medicine administered to an animal to expel parasitic worms.
*Vesicant* Substance producing blisters on skin or mucous membrane.
*Vestigeal* Small and rudimentary.
*Visceral* Relating to the internal organs of the body.

# BIBLIOGRAPHY

ALLEN, D.E. (1980) A possible scent different between *Crataegus species* Watsonia *13*:119–129.

ARY, S. and GREGORY, M. (1960) The Oxford Book of Wild Flowers. Oxford University Press.

AUTOMOBILE ASSOCIATION (1973) Book of the British Countryside. Edit. Drive Publications Ltd.

BALLANTYNE, G.H. (1984) Batology can be fun! (Study of brambles in Scotland with v/c table of distribution). BSBI Scott. News. 1.*6*:9–13.

BARTÍK, M. and PISKAČ, A. (1981) Veterinary Toxicology. Elsevier Scientific Publishing Co., Amsterdam.

BAXTER, C.P. (1983) Solomon's Seal poisoning in a dog. Vet. Rec. *113*:247–248.

BEAN, W.J. (1914) Trees and Shrubs Hardy in the British Isles. John Murray, London.

BELL, J.N.B. and TALLIS, J.H. (1973) Biological Flora of the British Isles. *Empetrum nigrum*.L. J.Ecol.*61*:289–305.

BIGNAL, E. (1980) The Endemic Whitebeams of North Arran, Glasg.Nat.*20*(part 1) 59–64.

BLACKMORE, S. (1983) Native Cherries of the British Isles. Living Countryside *9*:2153–2155.

BOWMAN, R.P. andCLEMENT, E.J. (1984) Blueberry, (*Vaccinium corymbosum*.L.) established in S.Hants. (V.C.11) BSBI News *No37*:20–21.

BROWICZ, K. (1968) Distribution of woody Rosaceae in west Asia, 2. On the distribution of *Mespilus germanica* L. Arb.Kórnieke Roczn., *13*:27–36.

BULL, A.L. (1981) The Brambles of Suffolk – further notes. Suffolk Nat.Hist.*18*:247–248.

BUTCHER, R.W. (1947) Biological Flora of the British Isles *Atropa belladonna* L.J.Ecol. *34*:345–353.

BUTCHER, R.W. (1961) A New Illustrated British Flora (2 vol). Leonard Hill (Books) Ltd, London.

CAMPBELL, M.S. (1945) The Flora of Uig (Lewis). T.Buncle & Co. Ltd., Arbroath.

CARVILL, P.H. (1980) Floristic notes from Co.Dublin (H.21) Bull.Irish Biogeogr.Soc.*4*:23–27.

CHANDLER, J.H. and GILBERT, J.L. (1970) *Crataegus monogyna* with yellow berries Watsonia *8*:85.

CLAPHAM, A.R. TUTIN, T.G. and WARBURG, E.F. (1981) Excursion Flora of the British Isles Edition 3. Cambridge University Press.

CLARKE, E.G.C. and CLARKE, M.L. (1975) Veterinary Toxicology. Baillière Tindall, London.

CLEMENT, E.J. (1978) Exotics 1977. *Viburnum rhytidophyllum* Hemsley in West Kent. Wild Flower Mag.*382*:23–25.

CLEMENT, E.J. (1978) Exotics 1977. *Rosa damascena* Mill. & *R.centifolia* L. from Cheddar. Wild Flower Mag.*382*:23–25.

CLEMENT, E.J. (1982) Alien Cotoneasters. Adventive News 23. BSBI News *32*:18–19.

CLEMENT, E.J. (1982) Alien Symphoricarpos. Adventive News 23. BSBI News *32*:20.

CLEMENT, E.J. (1983) Berry Catchfly *Cucubalus baccifer* L. in Britain. BSBI News *34*:34.

COOPER, M.R. and JOHNSON, A.W. (1984) Poisonous Plants in Britain and their effects on Animals and Man. H.M.S.O. Reference Book 161.

CULPEPER, N. (1982) Reprint of Culpeper's Complete Herbal. W.Foulsham & Co. Ltd., London.

DAVIES, J.A. (1983) *Cotoneaster integerrimus*: A step away from extinction? Welsh Bulletin *37*:10–11.

DAVIES, T.A.W. (1979) *Crataegus laevigata* in Pembroke. Nature in Wales *16*:213.

DEAKIN, R. (1871) The Flowering Plants of Tunbridge Wells and neighbourhood. Groombridge & Sons, London.

DEICHMANN, W.B. and GERARDE, H.W. (1964) Symptomatology & Therapy of Toxicological Emergencies. Academic Press.

DICKENS, C. (1850) The Miner's Daughters. Household Words *1*:127.

DONY, J.G. and DONY, C.M. (1986) Further notes on the flora of Bedfordshire. *Watsonia 16*:163–172.

DONY, J.G. ROB, C.M. and PERRING, F.H. (1974) English Names of Wild Flowers. Butterworth, London.

DUNN, M. (1984) Juniper survives in Balanaguard Glen. Scottish Wildlife *20*:21–24.

EDEES, E.S. (1983) *Rubus canterburiensis*. A new bramble from Kent. Watsonia *12*:279–280.

EDMONDS, J.M. (1977) Taxonomic Studies on *Solanum*

section. Solanum (Maurella). Bot.Journ.Linn.Soc.London *75*:141–178

EDMONDS, J.M. (1977) Black Nightshade Survey – interim report. BSBI News *No21*:12–13.

EDMONDS, J.M. (1979) A proposed Black Nightshade survey. Watsonia *12*:279–280.

EDMONDS, J.M. (1981) Black Nightshade Survey – third interim report. BSBI News *No28*:8–9.

ELLIS, R.G. (1983) Flowering Plants of Wales. National Museum of Wales.

EVANS, S.B. (1977) Plants of an oyster midden. Nature Wales *15*:141–142.

EVANS, T.G. (1979) Strawberry Tree in Wales. Nature Wales *16*:209.

FITTER, R.S.R. (1967) The Penguin Dictionary of British Natural History. Penguin Books.

FITTER, R.S.R. (1971) Finding Wild Flowers. Collins, London

FITTER, R and FITTER, A. (1974) The Wild Flowers of Britain and Northern Europe. Collins, London

FLORA EUROPAEA (1964–1980) 5 Volumes. Edit. by Tutin, T.G., Heywood, V.H., Burges, N.A., Moore, D.M., Valentine, D.H., Walters, S.M. and Webb, D.A. Cambridge University Press.

FORSYTH, A.A. (1968) British poisonous Plants. Min.Agric.Fisheries and Food. Bulletin 161. H.M.S.O.

FULLER, R.M. and BOORMAN, L.A. (1977) The spread and development of *Rhododendron ponticum* L. on dunes at Winterton, Norfolk in comparison with invasion by *Hippophaë rhamnoides* L. at Saltfleetby, Lincolnshire. Biol.Conserv.*12*:83–94.

GARNETT, P.M. and SLEDGE, W.A. (1967) The distribution of *Actaea spicata* L. Yorkshire Naturalist 73–76.

GARRARD, I and STREETER, D. (1983) The Wild Flowers of the British Isles. Macmillan, London.

GENT, R. (1681) The Experienced Farrier or Farring Compleated. Richard Northcott, London.

GERARD, J. (1597) Gerard's Herball. Edit. Marcus Woodward (1927). Reprinted by Spring Books (1964).

GODWIN, H. (1943) Biological Flora of the British Isles. *Rhamnus catharticus* L. J.Ecol.*31*:66–76.

GODWIN H. (1943) Biological Flora of the British Isles. *Frangula alnus* Miller. J.Ecol.*31*:77–92.

GORER, R. (1980) Rich in flowers, foliage and fruit: the rowans and whitebeams. Country Life *168*(4338):1238–1239.

GORISSEN, H. and WESTHOFF, V. (1983) Verspreiding en oecologie nav de Grote bosaard bei *Fragaria moschata* Duch. Natuurhist. Maandblad *72*:203–206.

GRENFELL, A.L. (1983) Aliens and Adventives. More on *Solanaceae* in Britain. BSBI News *No35*:12–14.

GREY-WILSON, C. (1979) The Alpine Flowers of Britain and Europe. Collins, London.

GRIEVE, M. (1931) A Modern Herbal. Reprinted by Peregrine Books (1976).

HADFIELD, M. (1968) The true Service Tree. Gardeners Chronical *163*:(19) 14–15.

HALL, P.C. (1980) Sussex Plants Atlas. Brighton Borough Council, Booth Museum of Natural History.

HALLIDAY, G. (1978) Flowering Plants and Ferns of Cumbria. Centre for North-West Regional Studies, University of Lancaster.

HARDIN, J.W. and ARENA, J.M. (1974) Human Poisoning from Native and Cultivated Plants. Duke University Press, Durham, North Carolina.

HARDY, E. (1973) *Hippophaë rhamnoides* L. – attractiveness to birds. Watsonia *9*:269–270.

HARMES, P. (1983) A new polymorphism in *Arum maculatum* L. BSBI News *No33*:20.

HARTLEY, P.H.T. (1954) Wild fruits in the diet of British thrushes. A study in the ecology of closely related species. British Birds *47*:97–107.

HARVEY, J and COLIN-JONES, D.G. (1981) Mistletoe hepatitis. Br.Med.J.*282*:186–187.

HEATHCOTE, D. (1982) Another unusual spindle. Suffolk Nat. Hist. *18*:320.

HILLMAN, E.M. and WARREN, A. (1973) Survey of *Ruscus aculeatus* on Bookham Common. Lond.Nat.52:93–103.

HILLMAN, E.M. (1979) A study of *Ruscus aculeatus* on Bookham Common. Lond.Nat.*58*:44–55.

HULL, P and SMART, C.J.B. (1984) Variation in two *Sorbus* species endemic to the Isle of Arran, Scotland. *Ann.Bot. (Oxford)* *53*:641–648.

HYDE, E.M. (1982) An unusual spindle (*Euonymus europaeus*) in West Suffolk. Suffolk Nat.Hist.*18*:319–320.

INGRAM, R. and NOLTIE, H.J. (1981) The Flora of Angus. Dundee Museums and Art Galleries.

JORDAN, M. (1976) A Guide to Wild Plants. Millington Books, London

KAY, Q,O.N. (1985) Dioecy and Pollination in *Viscum album*. BSBI Annual Exhibition, November 1985.

KAY, Q.O.N. and PAGE, J. (1985) Dioecism and pollination in *Ruscus aculeatus*. Watsonia *15*:261–264.

KEBLE-MARTIN, W. (1965) The Concise British Flora in Colour. Ebury Press and Michael Joseph.

KENNETH, E.G. (1978) Cranberry in Knapdale. Glasgow Nat. *19*:426

KENWORTHY, J.B. (1976) John Anthony's Flora of Sutherland. Botanical Society of Edinburgh.

KLOET, S.P. vander. (1977) Evidence against autogamy in *Vaccinium uliginosum* L. Proc. Nova Scotia Inst.Sci.*28*:101–104.

KNIGHT, G.H. (1962) Ivy in Woods. Ann.Rep.Warwick N.H.S *24*:6–9.

KNIGHT, G.H. (1978) Spurge-laurel berries. Ann.Rep.Warwick N.H.S *8*:45.

KUGLER, H. (1981) Zur Bestänbung von *Bryonia dioica* Jacq. Ber.Deutsch.Bot.Ges. *94*:287-290.

KYLE, R. (1983) Poisoning of Sheep by Lords-and-Ladies. Vet.Rec.*113*:23.

LAMPE, K.F. and FAGERSTRÖM, R. (1968) Plant Toxicity and Dermatitis. Williams & Wilkins Co., Baltimore.

LAMPE, K.F. (1981) Common Poisons and Injurious Plants. U.S. Department of Health and Human Services HHS Publication No.(FDA) 81–7006.

LANG, D.C. (1983) The Wild Flower Finder's Calendar. Ebury Press, London

LAUNERT, E. (1981) The Hamlyn Guide to Edible and Medicinal Plants of Britain and Northern Europe. Hamlyn, London.

LESLIE, A.C. and J.F. (1978) *Fragaria moschata* Duchesne and other Strawberries. BSBI November Exhibition Meeting.

LESLIE, A.C. (1978) *Fragaria moschata* Duchesne and *Fragaria vesca* L. BSBI News *No.18*:27.

LESLIE, A.C. (1978) The occurrence of *Solanum nigrum* L.x *S.sarrachoides* Sendtn. in Britain. Watsonia *12*:29–32.

LESLIE, A.C. (1979) *Fragaria moschata* Duchesne and *Fragaria vesca* L. BSBI News *No.21*:7.

LESLIE, A.C. (1980) *Solanum nigrum* L. subsp.*schultesii* (Opiz) Wessely in Britain. BSBI News *No.24*:21–22.

LEYEL, C.F. (1961) Culpeper's English Physician and Complete Herbal. Arco Publications.

LLOYD, E.G. (1977) The Wild Service Tree, *Sorbus torminalis*, in Epping Forest. Lond.Nat.*56*:22–28.

LOSKE, K.-H (1981) Zum Vorkommen von Epiphyten (nicht parasitäre Pflanzenbesidler) auf Kopfbäumen. Natur und Heimat *41*:18–26.

LOUSLEY, J.E. (1961) The Status of *Cucubalus baccifer* L. in England. Proc. BSBI *4*(3):261–268.

LOUSLEY, J.E. Biological Card Index System. British Herbarium, British Museum (Natural History).

MABEY, R. (1972) Food for Free. Collins, London

MCALLISTER, H. (1980) The Problem of *Hedera hibernica* resolved. Ivy Exchange Newsletter *3*:26–27.

MCALLISTER, H. (1982) New Work on Ivies. International Dendrology Year Book 1982 p1–4.

MCBARRON, E.J. (1976) Medical and Veterinary Aspects of Plant Poisons in New South Wales. Dept. of Agriculture, New South Wales.

MCCLINTOCK, D. and FITTER, R.S.R. (1956) The Pocket Guide to Wild Flowers. Collins, London,

MCCLINTOCK, D. (1969) A new Amelanchier. Gard.Chron.*165* (24):11–12 and *165* (26):5.

MACINTYRE, D. (1978) *Fragaria moschata* Duch. and *F.vesca* L. BSBI News *No20*:25.

MALTO-BELIZ, J (1982) Some reflections on Mediterranean plants in Ireland. Journ.Life Sc.Roy. Dublin Soc.*3*:227–282.

MARTINDALE (1982) The Extra Pharmacopoeia (28th Edition) Ed.J.E.F.Reynolds. The Pharmaceutical Press, London.

MELLANBY, K. (1982) Hedges – habitat or history. Natural World (Autumn edition).

MELVILLE, R. (1975) *Rosa* L. pp212–227 in Hybridization and the Flora of the British Isles. Ed.C.A.Stace. Academic Press.

MICHAEL, P. (1980) All Good Things Around Us. Ernest Benn, London.

MORRIS, M. (1980) *Cotoneaster integerrimus* – a conservation exercise. Nature Wales *17*:19–22.

NELSON, E.C. (1980) Naturalized *Fuchsia* Populations in Ireland. BSBI News *No24*:28.

NELSON, E.C. (1981) The Nomenclature and History in Cultivation of the Irish Yew, *Taxus baccata* 'Fastigiata'. Glasra *5*:33–44.

NELSON, E.C. (1982) Historical records of Irish Ericaceae. Irish Nat.Journ.*20*:364–369.

NELSON, E.C. and SYNNOTT, D.M (1982) The Irish Willow and Irish Whitebeam. Int.Dendrol.Soc.Year Book 1981 pp112–114.

NEWTON. A. (1980) Progress in British *Rubus* Studies. Watsonia *13*:35–40.

NORTH, P.M. (1967) Poisonous plants and fungi in colour. Blandford Press, London.

O'KEEFE, P. (1984) Plant records from Waterford (H6) and South Tipperaray (H7). Ir.Nat.J.*21*:279–280.

OLNEY, P. (1966) Berries and Birds. Birds *1*:98–99.

O'MAHONY, A (1973) *Crataegus laevigata* (Poir.)DC. A first definite record for Ireland. Irish Nat.Journ.*17*:345–355.

O'MAHONY, T. (1977) Current taxonomic and distributional research on the genus *Rosa* in Ireland. Irish Biogeogr.Soc.Bull.*1*:41–44.
O'MOORE, L.B. (1955) Poisoning of Cattle with Lords-and-Ladies *Arum maculatum* Irish Vet.J.*9*:146–147.
PAGE, J. (1985) Botanical section report for 1984. Soc. Guernesiaise Rep.Trans. *21*:445–447.
PALMER, J.R. (1982) Huckleberry and other Nightshades at Dartford. BSBI News *No32*:16–17.
PALMER, J.R. (1983) A severe outbreak of St.Lucie Cherry in W.Kent. BSBI News *No33*:9.
PATON, A. (1984) *Sorbus domestica* in Wyre Forest. Wild Fl.Mag.*401*:19.
PEARSALL, W.H. (1971 revised ed.) Mountains and Moorlands. New Naturalist Series. Collins, London.
PEARSON, M.C. and ROGERS, J.A. (1962) Biological Flora of the British Isles: *Hippophaë rhamnoides* L. J.Ecol.*50*:501–513.
PERRING, F.H. (1972) Mistletoe Survey. Watsonia *9*:202–203.
PERRING, F.H. and WALTERS, S.M (1976) Atlas of the British Flora. E.P. Publishing Ltd., Wakefield, Yorkshire.
PETERKEN, G.F. and LLOYD, P.S. (1967) Biological Flora of the British Isles: *Ilex aquifolium* L. J.Ecol.*55*:841–858.
PHILP, E.G. (1982) Atlas of the Kent Flora. Kent Field Club.
(Author and date unknown) Pilze und Wildfrüchte. Verbraucher Dienst Informiert Nordwestdruck H.O. Persiehl, Hamburg.
PLANT GUIDE Western Michigan Poison Centre 1840 Wealthy S.E. Grand Rapids, Michigan 49506, USA.
POLLARD, E., HOOPER, M.D. and MOORE, N.W. (1974) Hedges. Collins, London.
POLUNIN, O and HUXLEY, A. (1965) Flowers of the Mediterranean. Chatto & Windus, London.
POLUNIN, O. (1969) Flowers of Europe. Oxford University Press.
PRIME, C.T. (1954) Biological Flora of the British Isles: *Arum neglectum* (Towns.) Ridley (*Arum italicum* var.*neglectum* Townsend; *Arum italicum* auct.angl.non Mill.). J.Ecol.*42*:241–248.
PRIME, C.T. (1960) Lords and Ladies. Collins, London.
RAVEN J. and WALTERS, M. (1956) Mountain Flowers. The New Naturalist. Collins, London.
RIBBONS, B.W. (1971) Alder Buckthorn in the upper Forth Valley. Glasgow Naturalist *18*:583–584.
RICH, T.C.G. and BAECKER, M. (1986) The distribution of *Sorbus lancastriensis* E.F.Warburg. Watsonia *16*:83–85.
RITCHIE, J.C. (1955) Biological Flora of the British Isles: *Vaccinium vitis-idaea* L. J.Ecol.*43*:701–708.
RITCHIE, J.C. (1956) Biological Flora of the British Isles: *Vaccinium myrtillus* L. J.Ecol.*44*:291–299.
ROE, R.G.B. (1981) The Flora of Somerset. Somerset Archaeological and Natural History Society, Taunton Castle, Taunton.
RONWEDER, O. and URMI, E. (1978) Centrospermen-studien, 10. Untersuchunger über den Bau der Blüten und Früchte von *Cucubalus baccifer* etc. Bot.Jahrb.*100*:1–25.
ROPER, P. (1979) Wild service tree survey. BSBI News *No23*:7.
ROPER, P. (1982) Article on *Sorbus torminalis*. Int.Dendrology Year Book.
ROPER, P. (1982) Raising Wild Service Trees. BSBI News *No30*:20.
ROPER, P. (1982) Wild Service and other trees from seed. BSBI News *No31*:27–28.
ROSE, F. (1981) The Wild Flower Key. Frederick Warne, London.
ROSE, P.Q. (1980) Ivies. Blandford Press, Dorset.
RUTHERFORD, A. (1976) Irish Ivy Survey. BSBI News *No13*:17–19.
RUTHERFORD, A. (1979) Irish Ivy Survey. BSBI News *No22*:8–9.
RUTHERFORD, A. (1984) Ivies Native and Introduced. BSBI News *No36*:12–14.
SALISBURY, E.J. (1942) The Reproductive Capacity of Plants. G.Bell & Sons, London.
SALISBURY, E.J. (1961) Weeds and Aliens. Collins, London.
SCOTT, W.A. (1976) Alien Cotoneaster in Lanark. Glasgow Nat. *19*:340.
SEALY, J.R. and WEBB, D.A. (1950) Biological Flora of the British Isles: *Arbutus unedo* L. J.Ecol.*38*:223–236.
SHAW, J.M. (1986) Suspected cyanide poisoning in two goats caused by ingestion of crab apple leaves and fruit. Vet. Rec. *119*:242–243.
SILVERSIDE, A.J. (1981) Report of field outing to Ullapool, W.Ross 28th July–3rd August. Watsonia *13*:260.
SIMPSON, F.W. (1982) Simpson's Flora of Suffolk. Suffolk Naturalists' Society.
SITWELL. N. (1984) Britain's Threatened Wildlife. Collins/Threshold Books, London.
SMITH, M.E. (1979) And then there were four. The wild *Cotoneaster* in Britain. Country-life

166(4288):788.
STACE, C.A. (1975) Hybridization and the Flora of the British Isles (Edit). Academic Press, London.
STEWART, O.M. (1980) *Rosa arvensis* in Scotland. BSBI Scot.News *2*:7.
STIRLING, A. MCG. (1976) *Sorbus rupicola*. Distribution of *Sorbus rupicola* in Southern Scotland is reviewed. Glasgow Nat.*19*:340–341.
STIRLING, A. MCG. (1980) A provisional list of the Dunbartonshire v/c99. Brambles. Glasgow Nat.*19*:512–513.
SWANN, E.L. (1970) The Status of Berry Campion. Trans. Norfolk Norwich Nat.Soc.*21*:378–379.
SWENGLEY, N. (1976) The romance of the wild service tree. Ecologist *6*:228–229.
SYNNOTT, D.M. (1978) The status of *Crataegus laevigata* in Ireland. Glasra *2*:49–55.
TAMPION. J. (1977) Dangerous Plants. David & Charles, Newton Abbot.
TAYLOR, K (1971) Biological Flora of the British Isles: *Rubus chamaemorus* L. J.Ecol.*59*:293–306.
VAUGHAN, I.M. (1982) An introductory note on the native roses of Suffolk. Suffolk Nat.Hist.*18*:304–307.
VAZART, B. (1959) Biologie florale de l'Asperge *Asparagus officinalis* L. Rev.Gen.Bot.*66*:405–418.
VICKERY, A.R. (1979) Holy Thorn of Glastonbury. West Country Folklore No.12. J.Stevens Cox. The Toucan Press, St.Peter Port, Guernsey.
VICKERY, A.R. (1981) Traditional Uses and Folklore of *Hypericum* in the British Isles. Economic Botany 35(3):289–295.
WARD, L.K. (1981) The demography, fauna and conservation of *Juniperus communis* in Britain. The Biological Aspects of Rare Plant Conservation H.Synge (Edit.) pp319–329.
WARD, L.K. (1982) The Conservation of Juniper; longevity and old age. Journ.Appl.Ecol.*19*:917–928.
WAY, J.M. and CAMMELL, M.E. (1982) The distribution and abundance of the Spindle-tree (*Euonymus europeaeus*) in southern England. Journ.Appl.Ecol.*19*:929–940.
WEYDAHL, E.M. (1975) *Rubus chamaemorus* på Kvithamar. Medd. Det.Norske Myrselskap *3*:1–8.
WHUR, P. (1986) White bryony poisoning in a dog. Vet.Rec.*119*:411.
WILMOTT, A. (1977) *Sorbus torminalis* (L.) Crantz. in Derbyshire Watsonia *11*:339–344.
WILSON, M.A. (1967) Saving the Plymouth Pear. J. Devon Trust Nat.Conserv.*1*:481–482.
WOODVILLE, W. (1790) Medical Botany (engravings by James Sowerby).

# · INDEX ·

*Note:* page numbers in italics refer to line drawings; page numbers in bold type refer to colour plates.

*Abbreviations*
A = Group A (see p.13)
B = Group B (see p.13)
C = Group C (see p.13)
dist. = distribution
col. = ecology
fl. = leaf and flower
fr. = fruit
iden. = identification
p = poisonous
p? = possibly poisonous
ssp. = subspecies
tox. = toxicology
var. = variety/ies